Institute of Mathematical Statistics

LECTURE NOTES–MONOGRAPH SERIES

Volume 18

Selected Proceedings of the Sheffield Symposium on Applied Probability

Ishwar V. Basawa and Robert L. Taylor, Editors

University of Georgia

Institute of Mathematical Statistics
Hayward, California

The production of the *IMS Lecture Notes–Monograph Series* is managed by the IMS Business Office: Jessica Utts, IMS Treasurer, and Jose L. Gonzalez, IMS Business Manager.

Library of Congress Catalog Card Number: 91-77908

International Standard Book Number 0-940600-25-0

Printed in the United States of America

SYMPOSIUM ON APPLIED PROBABILITY,
SHEFFIELD ENGLAND, AUGUST 16-19, 1989

Back Row: David G. Kendall, D. R. Grey, Chris Cannings, J. D. Biggins, R. T. Durrett, Robert M. Loynes.

Front Row: Robert L. Taylor, Chris C. Heyde, Joe M. Gani, N. U. Prabhu, C. W. Anderson.

PREFACE

The 211th IMS meeting was a special Symposium on Applied Probability and was cosponsored by the University of Sheffield. The purpose of the Symposium was to focus attention on some of the more prominent directions in applied probability. The Symposium was part of an IMS initiative to increase involvement in probability and its applications, and it also served as an opportunity to acknowledge the many contributions of Joe Gani to applied probability and the profession. Consequently, it was very appropriate that the Symposium was held at The University of Sheffield.

The program co-chairs, Chris Cannings and N. U. Prabhu, are commended for organizing an excellent program involving 64 speakers and over 150 participants. Particular details on the program and speakers appeared in several issues of the IMS BULLETIN during 1989 & 1990. This volume of selected proceedings of some of the papers from the Symposium represents the quality and scope of the presentations. These papers were carefully refereed, revised and edited. Eleven of the 20 papers in this volume are from invited speakers of the Symposium.

The Symposium was supported by British Telecom plc, the Royal Society, the Science and Engineering Research Council, U.S. (London) Army Research Office and U.S. (London) Office of Naval Research Grant N00014-89-J-9042, U.S. Army Research Office Grant DAAL03-89-0099, U.S. National Science Foundation Grant DMS-8906957, and the U.S. Office of Naval Research Grant N00014-89-J-1830. These funds helped defray expenses for the participants, especially young scientists and invited speakers, and for the host institution, The University of Sheffield.

The editors are most grateful to the referees who are listed on page 278 for their careful and punctual work with respect to the submitted manuscripts. Sankar Sethuraman and Dharma Thiruvaiyaru worked very diligently in proofreading the retyped and edited manuscripts for accuracy. Special thanks go to Connie Doss and Victor Grubbs who were responsible for the final typing and electronic transfers of manuscripts.

Finally, support in the preparation of this volume by the Department of Statistics at the University of Georgia is gratefully acknowledged.

I. V. Basawa and R. L. Taylor
University of Georgia
Athens, Georgia
August, 1990

CONTRIBUTORS

Andre Adler, Department of Mathematics, Illinois Institute of Technology, Chicago, Illinois 60616 USA

David Aldous, Department of Statistics, University of California at Berkeley, Berkeley, California 94720 USA

Ishwar V. Basawa, Department of Statistics, University of Georgia, Athens, Georgia 30602 USA

Niels G. Becker, Department of Statistics, LaTrobe University, Bundoora, Victoria, AUSTRALIA 3083

J. D. Biggins, Department of Probability & Statistics, University of Sheffield, Sheffield, S3 7RH, England, UNITED KINGDOM

Erwin Bolthausen, Institute fur Angewandte Mathematik, Universitat Zurich, Ramistr. 74 CH-8001, Zurich SWITZERLAND

David Branson, Department of Mathematics, University of Essex, Wivenhoe Park, Colchester, CO43SQ, England, UNITED KINGDOM

Chris Cannings, Department of Probability & Statistics, University of Sheffield, Sheffield, S3 7RH, England, UNITED KINGDOM

Henry E. Daniels, Statistical Laboratory, Cambridge University, 16 Mill Lane, Cambridge CB2 1SB UNITED KINGDOM

Joe Gani, Department of Statistics & Applied Probability, University of California at Santa Barbara, Santa Barbara, California 93106 USA

P. E. Greenwood, Department of Mathematics, 121-1984 Mathematics Road, University of British Columbia, Vancouver, B.C., CANADA V6T 1Y4

D. R. Grey, Department of Probability & Statistics, University of Sheffield, Sheffield, S3 7RH, England, UNITED KINGDOM

Robert C. Griffiths, Department of Mathematics, Monash University, Clayton, Victoria, 3168 AUSTRALIA

Chris C. Heyde, Statistics Research Section, School of Mathematical Sciences, Australian National University, GPO Box 4, Canberra ACT 2601, AUSTRALIA

Kenneth J. Hochberg, Department of Mathematics and Computer Science, Bar-Ilan University, 52900 Ramat-Gan ISRAEL

Peter Jagers, Chalmers University of Technology & Goteburg University, School of Mathematical & Computing Sciences, S-412 96, Goteborg SWEDEN

John E. Kolassa, IBM Corporation, Thomas J. Watson Research Center, P.O. Box 218, Yorktown Heights, NY 10598 USA

Joop Mijnheer, Department of Mathematics, Leiden University, P.O. Box 9512, 2300 RA Leiden THE NETHERLANDS

Olle Nerman, Chalmers University of Technology & Goteburg University, School of Mathematical & Computing Sciences, S-412 96, Goteborg SWEDEN

Mina Ossiander, Department of Mathematics, Oregon State University, Corvallis, Oregen 97331-4605 USA

Michael Phelan, Department of Civil Engineering & Operations Research, Princeton University, Princeton, N.J. 08544 USA

Stanley A. Sawyer, Department of Mathematics, Campus Box 1146, Washington University, St Louis, Missouri 63130 USA

Ziad Taib, Chalmers University of Technology & Goteburg University, School of Mathematical & Computing Sciences, S-412 96, Goteborg SWEDEN

Robert L. Taylor, Department of Statistics, University of Georgia, Athens, Georgia 30602 USA

G. T. Vickers, Department of Probability & Statistics, University of Sheffield, Sheffield, S3 7RH, England, UNITED KINGDOM

Peter Whittle, Statistical Laboratory, Cambridge University, 16 Mill Lane, Cambridge, CB2 1SB UNITED KINGDOM

Lu Zhunwei, Department of Probability & Statistics, University of Sheffield, Sheffield, S3 7RH, England, UNITED KINGDOM

CONTENTS

INTRODUCTION

SOME ASPECTS OF THE SHEFFIELD SYMPOSIUM

Ishwar V. Basawa & Robert L. Taylor
University Of Georgia

For several years the Institute of Mathematical Statistics (IMS) has attempted to focus special and extra attention on some of the more prominent directions in probability and its applications. In accordance with this initiative the Symposium on Applied Probability at Sheffield in August 1989 was sponsored jointly by the IMS and the University of Sheffield.

This volume contains several major papers of current interest in the important area of applied probability. The research topics covered include models in epidemiology, genetics, random fields, branching processes, random walks, directed polymers and evolution time-scales among others. The readers will find a broad array of interesting problems discussed by eminent scientists in these exciting fields of applied probability.

This volume contains 18 technical articles on various topics of current interest, in addition to the introductory articles by Joe and Chris on some of the historical background and the current and future developments in applied probability. A brief outline of the contents follows. Henry E. Daniels discusses perturbation approximations for epidemics, and Peter Whittle introduces processes with history-dependent transition intensities. Peter Jagers et al give an account of their thoughts on the time scale of evolution under an attractive title: "When did Joe's great ... grandfather live?" David Aldous studies random walk on finite graphs whereas J. D. Biggins discusses martingales in branching random walks. Niels Becker is concerned with the analysis of infectious disease data via martingales. D. R. Grey and Lu Zhunwei consider extinction probabilities in branching processes with random environments, and Andre Adler discusses accumulation points in a random walk. P. E. Greenwood and Mina Ossiander give a scholarly account of a central limit theorem for evolving random fields. Erwin Bolthausen discusses directed polymers in a random environment. Michael Phelan discusses the use of point processes in modeling and inference for random fields for rainfall. Kenneth Hochberg gives an interesting account of measure-valued processes. John Kolassa and Joop Mijnheer discuss saddle point approximations and double stable integrals, respectively. The remaining four papers are on various aspects of models in genetics. David Branson studies an urn model for certain genetic processes, and Robert Griffiths discusses the ancestral graphs. Chris Cannings and G. T. Vickers give an account of genealogy patterns. Finally, Stanley Sawyer studies gene-conversion.

The excellent program of the Symposium, as represented by the selected papers in this volume, attracted many serious researchers to this Symposium which totalled over 150 participants. Another reason for the success of the Sheffield Symposium was that it is well recognized that Sheffield served as a major international center for the development of applied probability as a research area during 1965-1974 under the direction of Joe Gani. Furthermore, Sheffield continues to host the journals published by the Applied Probability Trust. During the symposium, a special banquet was held in Joe's honor. In his talk at the banquet, Joe gave an illuminating account of some of his efforts in developing the Department at Sheffield, and Chris Heyde delivered an address highlighting some of the many achievements of Joe Gani. Both Joe and Chris agreed to our request to contribute to this volume, and their articles which follow form an important part of this Introduction.

SOME REMARKS ON RECENT DEVELOPMENTS IN APPLIED PROBABILITY

J. Gani
University of California, Santa Barbara

Applied probability has finally achieved the recognition it deserves. The recent 211th Meeting of the Institute of Mathematical Statistics (IMS) in Sheffield, England, between August 16 and 19, 1989, stands out as a landmark in its acceptance as a well defined field of mathematical research.

There have been conferences specifically dedicated to applied probability before: possibly the earliest was the Seventh Symposium in Applied Mathematics, jointly sponsored in April 1955 at the Polytechnic Institute of Brooklyn by the American Mathematical Society and the Office of Ordnance Research. It was here that the term *Applied Probability* was first coined to describe the proceedings of speakers on Mathematical Probability and its Applications [1]. Not that applications of probability had been lacking previously. In June 1949, for example, at the Royal Statistical Society's Symposium on Stochastic Processes [2], in London, three pioneering reviews of the application of stochastic processes to statistical physics, evolutionary processes and population growth had been presented by Jo Moyal, Maurice Bartlett and David Kendall respectively. But the concept of applied probability, as a discipline somewhat distinct from probability theory, does not appear to have crystallized until 1955.

The 1989 Sheffield symposium in Applied Probability, coming soon after the IMS announcement of its new journal, the *Annals of Applied Probability* due to begin publication in 1991, marks a new chapter in the history of applied probability. Any remaining doubts among fellow professionals as to the legitimacy of the discipline will have been laid to rest. Applied probability may now be regarded as an area of importance commensurate with that of the overlapping fields of probability theory and mathematical statistics. Some of us had adopted this position many years ago, but our perception was far from universal.

The main themes of the initial symposium in *Applied Probability*, held in Brooklyn in 1955 were related to the physical sciences: the theory of diffusion, the phenomenon of turbulence, and the uses of probability in classical and modern physics. The participants included a highly distinguished group of probabilists and applied mathematicians who gave lectures on the following topics:

Paul Levy on Brownian motion depending on n parameters,

Joe Doob on the first-boundary problem,

Willy Feller on boundaries defined by stochastic matrices,

Eberhard Hopf on the statistical theory of turbulence,

Guido Münch on stochastic processes in astronomy,

George Batchelor on homogeneous turbulence,

Mark Kac on probability in classical physics,

Stan Ulam on infinite models in physics, and

B. O. Koopman on quantum and the foundations of probability.

Remarkably enough, this very promising start does not appear to have been followed up in the U.S.A. In Britain, however, Maurice Bartlett must have shared the symposium organizers' views: in the late 1950s, he named the Methuen series of which he was the general editor *Monographs in Statistics and Applied Probability.* By 1963, research in applied probability had intensified, and it became clear that a new publishing outlet was essential for the work of researchers who were exploring applications of probability to problems in the biological, physical, social and technological sciences.

In 1964, there was a sufficient number of papers being written in applied probability to justify my launching the *Journal of Applied Probability (JAP)*, a periodical jointly sponsored by the Applied Probability Trust (a foundation) and the London Mathematical Society. The distribution of JAP, which initially totalled roughly 250 copies per issue, rose rapidly to about 1500. By 1969, the number of submissions had grown large enough to warrant the publication of a companion journal, *Advances in Applied Probability (AAP).*

Interest in *JAP* in 1964 was greater among researchers in the U.K. and Australasia (together 48%), with their strong tradition of applied mathematics, than among those in North America (42%). But these relative positions have gradually altered: in the three years 1986-88, the majority of papers from the 41 countries represented in *JAP* and *AAP* (see regions in Table 1) originated from authors in North America (42%), followed closely by those in Eastern and Western Europe (together 30%), with UK and Australasian authors (17%) in third place.

Table 1

Numbers of papers in *JAP* and *AAP* with originating region of authors

	JAP 1964		*JAP/AAP* 1986-88	
	No.	%	No.*	%
USA and Canada	10.5	42.00	187	41.93
UK	2.5	10.00	51	11.43
Western Europe			115	25.78
Eastern Europe (incl. USSR)	1	4.00	21	4.71
Japan	1	4.00	15	3.36
Australia and New Zealand	9.5	38.00	23	5.16
Middle East, Africa, Asia	0.5	2.00	30	6.73
South and Central America			4	0.90
TOTAL	25.0	100.00	446	100.00

* The numbers in the third column contain minor inaccuracies due to rounding off errors when a paper has multiple authors. When a paper has two authors only, each gets credited with 0.5 of a paper.

During the 1970s, several new journals appeared which published material in applied probability: the *Annals of Probability* (1973), *Stochastics* (1973), *Stochastic Processes and their Applications* (1973), *Mathematics of Operations Research* (1976), *Stochastic Models* (1985), *Applied Stochastic Models and Data Analysis* (1985), and *Queueing Systems* (1986), to name but a few. The forthcoming *Annals of Applied Probability* demonstrates the IMS's strong commitment to the field, and will be welcomed by all applied probabilists.

Concern has been expressed that this new journal may attract some papers away from existing periodicals. I believe its more enduring effect will be to focus the efforts of young researchers on the varied and fascinating problems of applied probability arising in the biological, physical, social and technological sciences. I should be very surprised if the number of publications in applied probability did not grow substantially during the 1990s. The present journals may even find that they experience difficulties in accommodating all acceptable submissions, and I would expect some new journals to be started in the more specialized areas of the subject.

It is now possible, with hindsight, to trace the emergence of applied probability over the past 40 years; but its relevance to probabilistic and statistical studies was not always so readily accepted, even 15 years ago. This was apparent at the Conference on Directions for Mathematical Statistics [3] organized in Edmonton, Alberta, in August 1984 by Sudhish Ghurye. The conference was intended to discuss "... areas of current interest in which statistical methodology will yield substantive dividends."

Four of the twelve papers at the conference considered problems and issues in applied probability. Mark Kac in his paper 'Some reflections of a mathematician on the nature and role of statistics', spoke of applications of probability to physics, and to radar observations. He outlined Maxwell's statistical approach to problems in thermodynamics, and mentioned Siegert's work on the observation of radar signals during World War II. Ron Pyke, as the first editor of *Annals of Probability* wishing to encourage submissions in applied probability, reflected humorously on 'Applied probability: an editor's dilemma'. I took the opportunity to survey 'Theory and practice in applied probability', while Wendell Fleming gave an overview of 'Diffusion processes in population biology'.

Yet despite the unequivocal evidence placed before them, many of our statistical colleagues continued to treat applied probability as a field of marginal importance for a further decade. This seems to be the fate of many interdisciplinary areas in their earlier stages of development: biochemists have assured me that their difficulties have in many ways paralleled those experienced by applied probabilists. It is perhaps an unavoidable stage in the establishment of all new fields.

Has the scope of applied probability altered since 1955, when most of its problems were inspired by the physical sciences? A survey of published papers indicated that the areas of application have broadened steadily over the past 35 years. In 1971-3 for example, the main topics covered in *JAP* and *AAP*, ranked roughly in terms of decreasing numbers of papers, were stochastic models (branching processes, Markov chains and processes, other processes), queueing, storage and traffic theory, operations research, population processes in biology, epidemiology and mathematical genetics, distribution theory and limit theorems, and geometric probability. There were regrettably few contributions to the physical and social sciences.

Fifteen years later, in 1986-8, while the mix remained similar, there had been some new departures and substantial changes of emphasis: the number of biological papers had suffered a sharp reduction as more specialist journals in mathematical biology were published, but there had been an increase in the technological contributions, especially in queueing theory and networks, control and reliability. Roughly half the papers were on topics in probability theory relevant to applications, while the other half were concerned with the applications themselves. Only 4 papers were inspired by the physical and social sciences, most such contributions having found their way to specialist journals.

The 1989 Sheffield symposium in Applied Probability covered an exceptionally wide range of interests. There were over 60 invited and contributed papers on

Probability problems inspired by physics,

Probability problems inspired by biology, including mathematical genetics,

Models in epidemiology,

Probability in the theory of algorithms,

Extreme value theory,

Probability problems of queueing, including queueing networks,

Mathematical models of finance,

Point processes: applications and inference,

Self-similar processes,

Random processes and fields subject to long-range dependence,

Branching processes, and

The metric structure of Euclidean shape spaces,

among other topics. It was a source of great personal pleasure to me that David Kendall, who had contributed to the 1949 Royal Statistical Society Symposium and had been one of the strongest supporters of *JAP* in 1964, spoke on the last topic.

What of future directions of research in applied probability? If I were to hazard any predictions on the most exciting areas of investigation, I should pick algorithmic theory and methods, geometric probability, image processing and inference on random processes. Problems in these areas will continue to arise in the analysis of the idealized stochastic models of biology, genetics, epidemiology, physics and engineering, which scientists are so skilled in building. The real world is rich in random events whose study cries out for the skills of applied probabilists: the problems they pose will continue to challenge researchers for many years to come. I therefore feel I am on safe ground in forecasting an increasingly important role for applied probability in the 21st century.

I have no fear whatever that the sources of problems in applied probability will dry up; but I view the future production of mathematicians, statisticians and probabilists in our society with far less confidence. Steady reductions in the numbers of students in these disciplines have been reported, particularly at the graduate level, in North America, Europe, Britain and Australia. These decreases jeopardize not only the future of our profession, but more generally that of our society, based as it is on science and technology. We must respond to this threat with the utmost vigour; not only must we encourage more young students to graduate in the mathematical sciences, but we must also alert the leaders of our society and our politicians to the dangers of subjecting education too freely to the forces of a market economy. In science and technology, it is often the long term that counts. We must stress this viewpoint with persistence, patience and a wealth of convincing historical examples, if we wish our society -- and applied probability within it -- to develop healthily in future.

References

[1] McColl, L. A. (Editor) (1957) *Applied Probability; Proc. Seventh Symp. Appl. Maths Amer. Math., Soc.*, Vol. 7, McGraw-Hill, New York.

[2] Symposium on Stochastic Processes (1949), *J. R. Statist. Soc.* **B11**, 150-282.

Moyal, J. E.: Stochastic processes and statistical physics, 150-210.
Bartlett, M. S.: Some evolutionary stochastic processes, 211-229.
Kendall, D. G.: Stochastic processes and population growth, 230-264.
Discussion, 265-282.

[3] Ghurye, S. G. (Editor) (1975) *Proc. Conf. Directions for Mathematical Statistics*. Special Supplement *Adv. Appl. Prob.*, Applied Probability Trust, Sheffield.

THE PROMOTION AND DEVELOPMENT OF APPLIED PROBABILITY: A NOTE ON THE CONTRIBUTIONS OF J. GANI

C. C. Heyde
Australian National University

In his article in this volume (Gani (1990)), Joe Gani (JG) has given an outline of the genesis of an identity for Applied Probability. His modesty, however, has precluded him from indicating just how important his own role has been in the shaping of this identity. He has mentioned the coining of the term *Applied Probability* for an American Mathematical Society Symposium in 1955 but he has not mentioned his many publications, beginning from 1955, which in no small measure helped to establish the vitality of the newly identified sub-discipline.

This work, initially in dam and storage theory and Markov chains, but later in mathematical genetics, biology and epidemiology, saw the firm establishment of the new principle of specialization in applications of probability, as distinct from occasional forays into the area which had been common.

The time was ripe for such developments. The success of operations research during World War II was a precursor. The early 1950s had seen key publications in storage from Hurst (1951) and Moran (1954), in queues from Kendall (1951) and Lindley (1952) and in inventories from Arrow, Harris and Marschak (1951) and Dvoretzky, Kiefer and Wolfowitz (1952). Through these works a convenient mathematical foundation was laid down for topics which were to be central in applied probability. JG began his Ph.D. studies at the Australian National University under Moran in 1954 in time to ride the ensuing wave of activity. A list of his publications appears in Heyde (1988, pp. 1-13). Amongst the most influential of his early ones is Gani (1957) which provided a comprehensive review and synthesis of the current state of dam and storage theory and which stimulated much subsequent research.

The new activity in probability soon raised publication problems. Mathematical journals tended to favour only the more theoretical of the probability papers and statistical journals favoured applications directly concerned with statistics. Much good probabilistic research was not adequately catered for and the need for new avenues of publication became clear.

The first probability journal to be established was *Teoriya Veroyatnostei i ee Primeneniya* which appeared in 1956 and this was followed by *Zeitschrift für*

Wahrscheinlichkeitstheorie und verwandte Gebeite in 1962. Both made provision for applications, and were very influential from the outset, but in essence they were devoted to theory.

JG embarked upon the task of creating a journal to cater properly for work in applied probability in 1962. His labours came to fruition in 1964 with the creation of the Applied Probability Trust (APT) and the establishment of the *Journal of Applied Probability. Advances in Applied Probability* was added in 1969. The story of these beginnings is told in Gani (1982) and the preface to Gani (1988a) and it is a fascinating one of enterprise in the face of conservatism.

The *Applied Probability* journals have now been operating for more than a quarter of a century and they have had a profound influence on the evolution of probability and its applications. The richness of the subject and the symbiotic relationship between theory and applications are now well established. The spectrum of APT activities is, I believe, the single most significant of JG's achievements and it is the one which has given him the most persistent satisfaction (Gani (1982)).

Considerable autobiographical information appears in Gani (1988b) and a brief biography and catalogue of JG's achievements can be found in the preface to Heyde (1988). The *Applied Probability* journals will be an enduring legacy. However, his remarkable success in building a major Department of Probability and Statistics in Sheffield over the years 1965-74 and later the Manchester-Sheffield School of Probability and Statistics (1967-74) is particularly notable. Also in this category is the achievement of his period of seven years in CSIRO (1975-82) where he greatly expanded the scope of his Division from Mathematical Statistics to Mathematics and Statistics and changed its perspective to that of a high profile international operation. Government pressures and institutional restructuring later led to the Division going a different route but JG had shown what could be done and how it could be achieved. In all these developments, applied probability flourished as never before. It is in recognition of these achievements that the 1989 Sheffield Symposium in Applied Probability was held in his honour.

References

[1] Arrow, K. J., Harris, T. and Marschak, J. (1951) Optimal inventory policy, *Econometrika* **19**, 250-272.

[2] Dvoretzky, A. Keifer, J. and Wolfowitz, J. (1952) The inventory problem I, II, *Econometrika* **20**, 187-222, **20**, 450-466.

[3] Gani, J. (1957) Problems in the probability theory of storage systems, *J. R. Statist. Soc. B* **19**, 181-206.

[4] Gani, J. (1982) An adventure in publishing: eighteen years of the Applied Probability Trust, *Austral. J. Statist.* **24**, 1-17.

[5] Gani, J. (1988a) (Editor) *A Celebration of Applied Probability, J. Applied Prob.* Special Vol. **25A**.

[6] Gani, J. (1988b) Adventures in applied probability, *J. Applied Prob.* Special Vol. **25A**, 3-23.

[7] Gani, J. (1990) Some remarks on recent developments in applied probability, this volume.

[8] Heyde, C. C. (Editor) (1988). *Studies in Modelling and Statistical Science, Austral. J. Statist.* Special Vol. **30A**.

[9] Hurst, H. E. (1951) Long term storage capacity of reservoirs, *Trans. Amer. Soc. Civil Engrs.* **116**, 770.

[10] Kendall, D. G. (1951) Some problems in the theory of queues, *J. R. Statist. Soc. B* **13**, 151-173.

[11] Lindley, D. V. (1952) The theory of queues with a single server, *Proc. Camb. Phil. Soc.* **48**, 277-289.

[12] Moran, P. A. P. (1954) A probability theory of dams and storage systems, *Austral. J. Appl. Sci.* **5**, 116-124.

APPLICATIONS OF RANDOM WALKS ON FINITE GRAPHS

DAVID ALDOUS*
University of California, Berkeley

Abstract

This is a brief account of different contexts in which random walks on graphs occur, and of techniques used in their analysis. It is an expanded version of talks given in several places during 1986-89.

1. INTRODUCTION. Just as games of chance provide an easily comprehended setting for the elementary mathematics of probability, so random walks provide a simple illustration of stochastic processes. *Drunkard's walk*, with successive steps taken right or left at random, and *gambler's ruin*, the successive fortunes of a gambler staking $1 on red at roulette, are readily described to a non-mathematician. Generalizations to the study of sums of independent identically distributed random variables in 1 or d dimensions form a classic field of probability theory (usually called random walk, but here called *random flight*): see [40]. A different generalization is to consider a graph G and a particle stepping randomly around the vertices in the natural way: from a vertex v it chooses uniformly at random an edge e incident at v, and steps to the vertex v' at the other end of e. Such *random walks on graphs* are our subject.

The professional probabilist, vintage 1960-80, is liable to regard this topic as a dull special case of Markov theory. The author would prefer to draw an analogy with (say) branching processes, another special case of Markov theory which has long been recognized to have a distinctive flavor. The author, admittedly partisan, feels that "random walks on graphs" has at least as many applications, and a greater variety of mathematical tools and results, than does "branching processes", and deserves similar recognition as a subject in its own right.

Because these random walks arise in different settings, many elementary (and not-so-elementary) results have been repeatedly rediscovered. The purpose of this talk is to publicize the fact that a lot is known about these random walks, even though the subject has not been well-organized.

2. Contexts This section is devoted to a list of applications of random walks on graphs. "Applications" is a much-abused word: what I mean is "contexts in which random walks arise, in settings where the original question is not about random walks".

*Research supported by NSF Grant MCS 87-01426

Context 1. *Analogy with electric networks.* The delightful elementary monograph by Doyle and Snell [25] treats this analogy in detail. Consider the edges of a graph as a network of resistors of unit resistance. Apply unit voltage at vertex υ_1 and 0 voltage (ground) at vertex υ_1. Then some current $i_{\upsilon w}$ flows along each edge (υ,w), each vertex υ has some voltage $f(\upsilon)$ and some total current I flows through the network. These quantities are related to quantities defined in terms of the random walk on the graph. Write T_υ for first hitting time on υ, and T^+_υ for first return time to υ.

Write $E_\upsilon(\cdot)$ and $P_\upsilon(\cdot)$ for expectation, probability for the walk started at υ. Write $d(\upsilon)$ for the degree of υ (= number of edges at υ). Then

$$f(x) = P_x(T_{\upsilon_1} < T_{\upsilon_2})$$

$$I = d(\upsilon_1) P_{\upsilon_1}\left(T_{\upsilon_0} < T^+_{\upsilon_1}\right)$$

$$i_{\upsilon w} = E_{\upsilon_1}((\#\text{ steps } \upsilon \to w) - (\#\text{ steps } w \to \upsilon) \text{ before } T_{\upsilon_0})$$

Facts about electric networks, such as

- Removing a wire can only decrease current flow
- actual current flow minimizes energy dissipation amongst all possible

flows can be re-interpreted as facts abour random walks on graphs, and lead to results such as conditions for recurrence/transience for walks on infinite graphs.

Context 2. *Leveling networks.* Maps show heights of mountain-tops, etc., which are ultimately derived from measurements of height differences between points with a clear line of sight. Consider a graph whose vertices are the physical points whose heights are desired, and where an edge (υ,w) denotes that a measurement $a(\upsilon,w)$ of "height at υ - height at w" has been made. Let υ_0 be a reference point of known height $h(\upsilon_0)$. Assuming the measurements contain errors, an estimate of the true heights h(v) could be made by least squares: choose the function h which minimizes

$$\sum_{edges}\sum (h(\upsilon) - h(w) - a(\upsilon, w))^2$$

or more generally a weighted sum of this kind. It turns out that the least squares estimate $h^*(\upsilon)$ can be described in terms of the random walk X_n started at υ_0;

$$h^*(v) = E \sum_{n=1}^{T_v} a(X_n, X_{n-1})$$

In other words, each deterministic walk from v_0 to v yields an empirical height for v as the sum of measured height differences along that walk; and the least squares estimate is the average of these empirical heights, where we use the random walk to do the averaging. These ideas are explored in a somewhat neglected monograph by Borre and Meissl [13].

Context 3. *Recreational problems.* Imagine a knight on a corner square of an otherwise empty chessboard. Move the knight by choosing at random from the legal knight-moves. What is the mean time

- until the knight first returns to the starting square?
- until the knight has visited all squares?

These look like messy problems for which no simple answer would be expected. It turns out that Markov chain theory gives a simple answer (168) to the first problem (this is a textbook exercise). The second problem really is messy, but some asymptotics suggest a rough approximation (500) which simulations reveal to be about right.

Of course, the knight is performing random walk on a graph. Random walks arise in other recreational problems (on polyhedra; in card-shuffling), and from time to time appear in the "math" section of electronic bulletin boards. Diaconis [21] contains a nice account of card-shuffling problems.

Context 4. *Universal traversal sequences.* It has long been realized that probabilistic methods are sometimes useful in combinatorics to establish the existence of objects which are hard to exhibit constructively: classical applications are in the monograph by Erdos and Spencer [27]. The following elegant application of random walk to a combinatorial problem was made by Aleliunas et al [6]. Let $S(N,d)$ be the set of all d-regular graphs G with N vertices and with the edges at each vertex labelled $(1,2, \ldots,d)$. A *universal traversal sequence* $i_1, i_2, \ldots, i_U \in \{1, \ldots, d\}$ satisfies: for each $G \in S(N, d)$ the deterministic walk "at step n choose edge i_n" visits every vertex. What is the shortest length $U = U(N,d)$ of such a sequence? It can be proved that *most* sequences of length $O(dN^3 \log N)$ are universal traversal sequences.

To see why, choose $i_1, i_2, \ldots$ uniformly at random, so the walk is simple RW. On a fixed G, consider the *cover time* C = time to visit all vertices. It can be shown that $EC = O(N^2)$ for G regular. Then by iterating, there is a constant K such that on fixed G,

$$P(C > KdN^3 \log N) \le N^{-Nd}$$

But $|S(N,d)| = o(N^{Nd})$, so

$$C \le KdN^3 \log N \text{ for all } G \in S(N,d)) -.$$

Context 5. *Minimization algorithms.* One way in which probability enters the theory of algorithms is via randomized algorithms, as in the example above. Another way is that certain problems can be proved "hard" by showing that, when presented with random data, no algorithm has much chance of finding the solution quickly. An application of random walk was given by Aldous [3]. Consider a function h defined on the vertices of a graph G: for definiteness, the cube graph in d dimensions. One wants to find the vertex υ at which $h(\upsilon)$ is minimized. Constrain h to have no local minima except the global minimum. Thus any deterministic "descent" algorithm will work, but it might work slowly. Could there be some more sophisticated algorithm which always works quickly? The answer is no. Consider random walk started at a uniform vertex υ_0 and let h(v) be the first hitting time on υ. Then h is a random function satisfying the constraint, but it can be proved that *every* algorithm requires expected time $\Omega\left(2^{\frac{d}{2}-\varepsilon}\right)$ to locate υ_0.

Context 6. *Approximate counting, or simulating uniform distributions on combinatorial sets.* Consider finite combinatorial sets S_K, e.g. all permutations of 1,2,...,K satisfying some family of conditions. It is intuitively clear that finding an exact formula for $|S_K|$ is roughly equivalent to being able to write a program to simulate a uniform element of the set using a deterministic number of calls to a random number generator. But there is a classical "Markov chain" method of simulating approximately a desired distribution π, originally developed for Ising-type physics models [29]: define a chain whose stationary distribution is π and simulate the chain. In our setting, we make a graph G with vertex-set S_K and with edges (υ,w) interpreted as "υ can be obtained from w by a small change", for some specified notion of *small change.* If the graph is connected and regular (the non-regular case can be handled by minor modifications) then random walk has uniform stationary distribution, and so we can approximate the uniform distribution by simulating the random walk. For this method to be useful, the walk has to be "rapidly mixing": that is, the time for distributions to approach the stationary distribution should be small compared to the size of the graph. This is equivalent to G being an *expander* (i.e. highly- connected) graph.

Thus if S_K can be given a usable graph structure which gives an expander graph, we can simulate the uniform distribution approximately, and thus estimate quantities such as

$$h^* = \text{ave}_{s \in S} h(s)$$

for specified $h: S \to R$. A more interesting issue is to use such simulation to approximately count the size $|S_K|$. This is studied in detail by Sinclair and Jerrum [39], who develop a notion of "self-reducibility" for families (S_K) which permit approximate counting. To specify a simple example, for a given graph G let S_G be the set of partitions

$$s = \{ (v_1, w_1), (v_2, w_2), \ldots, (v_k, w_k) \} ; k \le |G|/2$$

into edges (v_i, w_i) where the vertices $v_1, w_1, v_2, w_2, \ldots$ are distinct. There is no usable exact formula for $|S_G|$, but we can make S_G into a graph with edges (s_1, s_2) such that the edges of G comprising s_2 and s_1 are identical except for some one edge (v_i, w_i). Performing random walk on S_G allows approximate counting of that set.

A continuous variant of the above is the problem of approximating the volume of a convex body $S \subseteq R^d$ in high dimensions d. By considering a fine grid, one can re-interpret this as the problem of approximate counting for a "convex" subset $S^* \subseteq Z^d$. One can then seek to do the counting by running random walk on S^*, considered as a subgraph of Z^d. This is studied in detail by Dyer et al [26].

Context 7. *Random spanning trees in regular graphs.* A connected finite graph G has a set of *spanning trees*, that is subgraphs on all vertices which are trees. The number of spanning trees is given by a classical expression, the *matrix-tree formula,* found in many textbooks on graph theory. Now consider the (uniform) random spanning tree in G. Studying its properties via combinatorial methods is not easy, since this involves counting the number of spanning trees with some extra property, and typically there are no nice formulas available. However, there is a probabilistic construction of the random spanning tree using random walk, as follows. Start the walk $X(n)$ from an arbitrary vertex $X(0)$. For each $v \neq X(0)$ let T_v be the first hitting time on v. Then the graph with edges $(X(T_v-1), v)$ is a random spanning tree, and it turns out (but seems non-obvious) that this random tree is indeed uniform on the set of all spanning trees. Thus one can use properties of random walk to study the random spanning tree. See Aldous [4] and Broder [16]}.

Context 8. *Planar graphs.* The usual square lattice in R^2 has edges of constant length and vertices of degree 4. The triangular lattice has edges of constant length and vertices of degree 6. There is a sense in which "6" is critical. Conklin [19], using results of Dodzuik [23] and Doyle and Snell [25], proves the following result.

Let G be an infinite planar graph with edge-lengths l_e. Suppose there exist $\varepsilon > 0$, $n_0 < \infty$ such that, for each connected finite set of vertices of size at least n_0, the average degree is at least $6 + \varepsilon$. Then $\sup_e l_e / \inf_e l_e = \infty$.

The proof can be phrased in terms of random walk. The "average degree > 6" hypothesis implies the walk is transient; whereas if G is planar and the edge lengths were bounded above and below then the walk would be recurrent.

Context 9. *Voter Models.* One of the major post-1970 developments in probability theory has been the study of interacting particle systems on infinite lattices: see the book of Liggett [35]. The following finite version has been studied by Donnelly and Welsh [24]. Consider a graph where each vertex is colored, initially with different colors. Each vertex from time to time (precisely, at times of independent Poisson processes) picks an adjacent vertex at random and changes its color to the color of the picked neighbor. Eventually, on a finite graph, all vertices will have the same color: how long does this take? This question turns out to be related (via a certain notion of *duality*) to the following question. Imagine particles, initially one at each vertex, which perform random walk on the graph, but which coalesce when they meet. Eventually they will all coalesce into one particle: how long does this take? Such questions are analogous to the cover time problems in Context 4.

These questions have been studied more in the setting of infinite graphs. There, the chance two walks coalesce may be less than 1, which turns out to be equivalent to the original voter model having more than one stationary distribution.

Context 10. *Diffusions on regular fractals.* The d-dimensional diffusion (= continuous-path, strong Markov) processes arising in most applications behave locally like Brownian motion with some drift and covariance functions, plus some boundary behavior. Quite different diffusions can be constructed to live on regular fractal sets such as the Sierpinsky gasket (divide an equilateral triangle into 4 smaller triangles, remove the center one, and repeat this procedure on the 3 remaining regions.) One can regard the fractal set S as a limit of finite graphs S_K and the diffusion on S as the renormalized limit of random walks on S_K. The properties of the diffusion may be established by first studying properties of the finite random walks. See Kusuoka [34], Barlow and Perkins [7], Krebs [33].

Context 11. *Random fractals.* Physicists believe that at "critical points" there is typically some type of "fractal" behavior. In undergraduate probability theory, the word *critical* arises in one place, in the context of simple branching processes. Consider a non-trivial critical (mean number of offspring = 1) branching process, and condition on non- extinction (i.e. condition on non-extinction at generation n and take weak limits of the whole process as $n \to \infty$.) This process, considered as a tree in the natural way, is a prototype for a class of "random fractals occuring at critical points in random processes". More complicated and more interesting random fractals occur as critical percolation clusters. One method of studying these fractal sets has been to study random walk on them. For instance, on the critical branching tree the distance from the origin at time n grows at rate $n^{1/3}$: see Kesten [32].

3. Topics and Techniques. We now turn from the extrinsic to the intrinsic. If we agree that random walks are worth studying, what are natural mathematical questions to ask about them? Let us list some topics which have been studied, and mention some techniques used. My own particular interest has been the case of large, finite, unstructured graphs, and that will be the central theme. But many of the topics discussed will be one step removed from this area, in one of many possible directions.

Technique 12. *Reversible Markov chains.* Simple random walk is an example of a *reversible* Markov chain. Where the graph has no special symmetry, there are few special techniques available beyond the general techniques for treating reversible chains. Thus it is arguable that the "natural mathematical setting" for the study of random walk is within the larger class of general reversible Markov chains. From this viewpoint, one special property of our random walks is that there is a simple explicit formula for the stationary distribution

$$\pi(v) = \frac{d(v)}{\Sigma_w d(w)}; d(v) = \text{degree of } v \qquad (1)$$

and that this stationary distribution is typically roughly uniform. This contrasts with other settings where reversible chains arise:

- queueing networks - Kelly [31]
- interacting particle systems - Liggett [35]
- physical systems - Whittle [43]

where useful explicit formulas are often hard to find, and where typically one gets stationary distributions of the form $\pi(x) = C\exp(-V(x))$.

Topic 13. *Exact formulas/ group representation theory.* Exact formulas for n-step distributions or first hitting times can only be expected for graphs with very special structure. Given a discrete group G and a subset H (typically, H is a small set of generators) with $H = H^{-1}$, we can construct the *Cayley graph* which has vertex-set G and edges $(g, gh) : g \in G, h \in H$. Such graphs satisfy one of a hierarchy of symmetry conditions (see e.g. Biggs [10].) Given enough symmetry, it becomes feasible to use group representation theory (the analog of Fourier analysis for random walk in R^d) to get more-or-less explicit formulas. Diaconis [21] gives a fine introduction.

Problem 14. *Weakening symmetry conditions.* For most applications it is unreasonable to assume any group-theoretic symmetry condition. On the other hand, if a graph is merely assumed to be regular, then the random walk may behave quite badly. As a specific instance, consider asymmetry of mean hitting times. Vertex-transitivity implies $E_v T_w = E_w T_v$, whereas for regular graphs the ratio may grow polynomially fast as the size of the graph increases. One would like some "niceness" condition which forced ratios $E_v T_w / E_w T_v$ to stay bounded or grow only slowly.

Technique 15. *Hard Analysis.* Random walks specialize reversible chains, which in turn may be considered as discretizations of d-dimensional reversible diffusions (without the geometry of d-dimensional space), which in turn generalize the physics of the heat equation and the mathematics of Brownian motion. This gives an indirect link to a large chunk of classical physics and analysis, and some of the ideas developed there carry over to our setting. One natural concept is the *Dirichlet form*

$$<f,f> \equiv \frac{1}{2|E|} \sum_{(i,j) \in E} (f(j) - f(i))^2 \tag{2}$$

where f is a function on the vertices and E denotes the set of edges. For instance, the dominant eigenvalue λ_2 controlling the asymptotic rate of convergence to the stationary distribution can be characterized as

$$\lambda_2 = 1 - \inf_f \frac{<f,f>}{\Sigma_i \pi(i) f^2(i)} \tag{3}$$

There are several ways to actually bound this quantity for specific graphs: an elegant way, involving specifying a path between each pair of vertices and seeking to minimize the number of such paths using any fixed edge, was introduced in [30] and is treated systematically by Diaconis and Stroock [22]. Deeper results relating the distributions at time t to properties of the Dirichlet form are given by Carlen et al. [17] in the more general setting of reversible chains.

Another powerful result whose proof comes from classical analysis is a universal large deviation bound due to Carne [18]. For random walk (X_n) on any regular graph,

$$P_v(X_n = w) \le P(|S_n| \ge d(v,w))$$

where S_n is simple symmetric random walk on the integers and d is graph-distance (= minimum path length). When $d(v,w) \gg n^{1/2}$, the elementary Binomial bounds make the right side small. In particular, for graphs of polynomial growth (i.e. the number of vertices within a ball of radius r grows only polynomially) we can deduce that $d(X_n, X_0) = O(n^{1/2} \log^a(n))$. Sharp results of this kind for subgraphs of Z^d are given by Barlow and Perkins [8].

Topic 16. *Infinite and finite graphs.* For random walk on an infinite graph, one can ask questions like: is it transient or recurrent? does it have non-trivial boundary? At first sight such questions have no relevance for finite graphs. But it turns out that they do relate to properties of *sequences* G_K of finite graphs with $|V_k| \to \infty$. Loosely, with an infinite graph G one can often associate finite graphs G_K which can be thought of as "finitizations" of G. For instance, with Z^d we associate Z^d modulo K. On the other hand there are many sequences of finite graphs (e.g. the cube graphs) with no natural limit infinite graph: in a sense, the notion of "sequence of finite graphs" is more general than "infinite graph".

For example, a sequence G_K of finite graphs may or may not have the property

$$\text{ave}_{vw} E_v T_w / |V_K| \text{ is bounded in K.} \tag{4}$$

When the G_K are finitizations of an infinite G, this property is equivalent to transience of the random walk on G. Thus in general we can regard this property as the finite analog of transience.

Technique 17. *Naive and threshold limits.* Classical probability theory was much concerned with the asymptotic behavior of a fixed process as time $\to \infty$. Such *naive limit theory* is rather trite for random walks on finite graphs. Instead, one can consider non-asymptotic properties of random walk as the size of the graph increases. Limit theorems in this setting are sometimes called *threshold* results, a term used for analogous results in the theory of random graphs (see Bollobas [12]}). As an example, under the boundedness assumption (4) one can show that normalized first hitting distributions converge to a limiting exponential distribution: see Aldous [1].

Topic 18. *Isoperimetric inequalities.* An inequality of the type: for all finite subsets A of vertices of a graph G,

$$\#\text{edges}\,(A \to A^c) \geq c \min\,(|A|, |A^c|)$$

is called an isoperimetric inequality. The largest c permitted (strictly positive, for a finite connected graph) is the isoperimetric constant for G. One can use c to bound λ_2 (recall (3)) and thence the time for convergence to the stationary distribution: see Jerrum and Sinclair [39] (this is one technique used in Context 6). Thus for finite G, the property that c is not close to 0 (such G's are loosely called *expanders*) is equivalent to the rapid mixing property of random walk on G.

For an infinite graph we have a rigorous definition: it is an expander iff c > 0. This graph-theoretic property corresponds sharply to some probabilistic properties of the random walk:

- The random walk has non-trivial boundary
- $P_v(X_n = v) \to 0$ exponentially fast.

See Gerl [28] for a survey. These results allow us to regard "rapid mixing" for walks on finite graphs as the finite analog of "non-trivial boundary" for walks on infinite graphs.

A "pure math" treatment of the connection between eigenvalues and expanders, in the context of graphs associated with groups, is in the forthcoming book by Lubotzky [36].

Technique 19. *Analytic and probabilistic parameters.* One can define parameters of a finite graph which measure typical values of quantities such as mean hitting times, cover times, or convergence times. Much of the theory of random walk deals explicitly or implicitly with inequalities between such parameters. It turns out there are two parallel theories for such parameters. One is an analytic theory, using eigenvalues such as λ_2, which can be interpretated as an "average-case" (over starting places) or L^2 theory. The other is a probabilistic theory, using techniques such as coupling (see e.g. Diaconis [21]; Aldous [5]}) to study the "worst-case" (over starting places) or L^∞ theory. The interplay between these two methods is a recurring theme.

Technique 20. *Pure graph theory.* The part of pure graph theory which relates most to our concerns is perhaps the spectral theory of graphs, treated in detail by Cvetkoviec et al [20]. [*Aside.* Actually [20] treat spectral theory of the adjacency matrix, which has no direct probabilistic relevance. More recently graph theorists have studied spectral theory of the Laplacian matrix Q, which is essentially the transition rate matrix for the continuous-time Markov chain with transition rate 1 along each edge: see e.g. Mohar [37]. Our concern is primarily with the transition matrix P, whose study reduces to that of Q iff the graph is regular. Of course, similar techniques can be used on all these matrices.]

It would be foolish to pretend that our random walk studies are central to pure graph theory. Many topics in graph theory (chromatic number, perfect graphs) involve qualitative or quantitative properties of graphs which do not seem to relate significantly to properties of random walk on the graph. Let us merely observe that *occasionally* random walk is useful in answering non-random questions about graphs (e.g. Contexts 4, 6, 8). And *occasionally* considering random walk on graphs will illuminate connections between properties of graphs. For example, the connection between eigenvalues and expanders is a little mysterious in itself, but become clear when you realize that both measure how rapidly random walk is converging to its stationary distribution.

Topic 21. *Exact extremal graphs.* The subject of extremal graphs - i.e. those for which general inequalities become equalities - is an established part of graph theory (see e.g. Bollobas [11]). As mentioned before, much of the theory of random walks involves bounds between parameters of the walk and parameters of the graph, so it is natural to ask for the exact extremal graphs where the bounds are attained. It is easy to show that the complete graph is extremal in various senses. For example, there is a general inequality

$$E_\pi T_\upsilon \geq (1-\pi(\upsilon))^2/\pi(\upsilon)$$

with equality iff the graph is complete. (π is the stationary distribution (1)). The only known hard extremal result about general graphs concerns $t^* \equiv \max_{\upsilon, w} T_\upsilon$. Brightwell and Winkler [15] show that t* is maximized, over n-vertex graphs, by a graph consisting of a complete graph on $k \approx 2n/3$ vertices, with an attached path of length n-k.

On *trees* there are simple explicit formulas (apparently due to Moon [38]) for means and variances of hitting times, so it is not surprising that extremal results are somewhat easier. Results concerning hitting times are given by Yaron [44], and results concerning cover times by Brightwell and Winkler [14]}.

Topic 22. *Fractional dimension.* For a subgraph G of the lattice Z^d, or more generally for a graph in which the number of vertices within a ball of radius r is $O(r^d)$ for some d, it is natural to try to assign some "fractional dimension" $\dim(G) \leq d$ to G. As mentioned in Context 11, one can try to relate definitions of dimension to properties of random walk on G : for recent work see Barlow and Taylor [9], Telcs [41, 42].

Technique 23. *Martingale methods.* Martingales occur naturally in some questions about random walk on *infinite* graphs, e.g. boundary theory. But the key results about martingales are the general inequalities, which are purely finitistic; one would expect these to be useful in establishing inequalities concerning finite random walks. Surprisingly, very little use of martingales is made in the existing literature (Aldous [2] is one exception): perhaps they will be exploited more in future.

A more extensive bibliography concerning random walks on finite graphs is available from the author.

References

[1] D. J. Aldous. Hitting times for random walks on vertex-transitive graphs. *Math. Proc. Cambridge Phil. Soc.*, 106:179-191, 1989.

[2] D. J. Aldous. Lower bounds for covering times for reversible Markov chains and random walks on graphs. *J. Theoretical Probab.*, 2:91-100, 1989.

[3] D. J. Aldous. Minimization algorithms and random walk on the d-cube. *Ann. Probab.*, 11:403-413, 1983.

[4] D. J. Aldous. *The Random Walk Construction of Uniform Spanning Trees and Uniform Labelled Trees.* SIAM. J. Discrete Math, 3: 450-465, 1990.

[5] D. J. Aldous. Random walks on finite groups and rapidly mixing Markov chains. In *Seminaire de Probabilities XVII*, pages 243-297, Springer-Verlag, 1983. Lecture Notes in math. 986.

[6] R. Aleliunas et al. Random walks, unviersal traversal sequences, and the complexity of maze traversal. In *Proc. 20th IEEE Found. Comp. Sci. Symp.*, pages 218-233, 1979.

[7] M. T. Barlow and E. A. Perkins. Brownian motion on the Sierpinski gasket. *Probab. Th. Rel. Fields*, 79:543-624, 1988.

[8] M. T. Barlow and E. A. Perkins. Symmetric Markov chains in Z^d: how fast can they move? *Probab. Th. Rel. Fields*, 82:95-108, 1989.

[9] M. T. Barlow and S. J. Taylor. *Fractional Dimension of Sets in Discrete Spaces*. Technical Report, Cambridge University, 1989.

[10] N. Biggs. *Algebraic Graph Theory.* Cambridge University Press, 1974.

[11] B. Bollobas. *Extremal Graph theory.* Academic Press, London, 1978.

[12] B. Bollobas. *Random Graphs.* Academic Press, London, 1985.

[13] K. Borre and P. Meissl. *Strength Analysis of Leveling-type Networks.* Geodaetisk Institut, Copenhagen, 1974. Volume 50.

[14] G. Brightwell and P. Winkler. *Extremal Cover Times for Random Walks on Trees*. Technical Report, Emory University, 1989.

[15] G. Brightwell and P. Winkler. Maximum Hitting time for Random Walks on Graphs. *Random structures and algorithms*, 1: 263 - 276, 1990.

[16] A. Broder. Generating random spanning trees. In *Symp. Foundations of Computer sci,*. pages 442-447, 1989.

[17] E. A. Carlen, S. Kusuoka, and D. W. Stroock. Upper bounds for symmetric Markov transition functions. *Ann. Inst. H. Poincare Sect. **B*** Suppl. 2:245-287, 1987.

[18] T. K. Carne. A transmutation formula for Markov chains. *Bull. Sci. Math. (2)*, 109:399-405, 1985.

[19] J. E. Conklin. *The Discrete Laplacian: Applications to Random Walk and Inverse Problems on Weighted Graphs*. PhD thesis, University of Rochester, 1988.

[20] D. M. Cvetkovic, M. Doob, and H. Sachs. *Spectra of Graphs*. Academic Press, 1980.

[21] P. Diaconis. *Group Representations in Probability and Statistics*. Institute of Mathematical Statistics, Hayward CA, 1988.

[22] P. Diaconis and D. Stroock. *Geometric Bounds for Eigenvalues of Markov Chains*. Technical Report, Harvard University, 1989.

[23] J. Dodziuk. Difference equations, isoperimetric inequality and transience of certain random walks. *Trans. Amer. Math. Soc.*, 284:787-794, 1984.

[24] P. Donnelly and D. Welsh. Finite particle systems and infection models. *Math. Proc. Cambridge Philos. Soc.*, 94:167-182, 1983.

[25] P. G. Doyle and J. L. Snell. *Random Walks and Electrical Networks*. Mathematical Association of America, Washington DC, 1984

[26] M. Dyer, A. Frieze, and Ravi Kannan. A random polynomial time algorithm for approximating the volume of convex bodies. In *Proc. 21st ACM Symposium on Theory of Computing*, pages 375-381, 1989.

[27] P. Erdos and J. Spencer. *Probabilistic Methods in Combinatorics*. Academic Press, 1974.

[28] P. Gerl. Random walks on graphs with a strong isoperimetric property. *J. Theoretical Probab.*, 1:171-188, 1988.

[29] J. M. Hammersley and D. C. Handscomb. *Monte Carlo Methods*. Methuen, 1964.

[30] M. Jerrum and A. Sinclair. Approximating the permanent. *SIAM J. Computing*, 18:1149-1178, 1989.

[31] F. P. Kelly. *Reversibility and Stochastic Networks*. Wiley, 1979.

[32] H. Kesten. Subdiffusive behavior of random walk on a random cluster. *Ann. Inst. H. Poincare Sect. B*, 22:425-487, 1987.

[33] W. B. Krebs. *A Diffusion Defined on a Fractal State Space*. PhD thesis, U. C. Berkeley, 1988.

[34] S. Kusuoka. A diffusion process on a fractal. In K. Ito and N. Ikeda, editors, *Probabilistic Methods in Mathematical Physics*, North-Holland, 1987.

[35] T. M. Liggett. *Interacting Particle Systems*. Springer-Verlag, 1985.

[36] A. Lubotzky. *Discrete Groups, Expanding Graphs and Invariant Measures*. SIAM, 1990. CMBS-NSF Regional conference Series, to appear.

[37] B. Mohar. *The Laplacian spectrum of Graphs*. Technical Report, University of Ljubljana, 1989.

[38] J. W. Moon. Random walks on random trees. *J. Austral. Math. Soc.*, 15:42-53, 1973.

[39] A Sinclair and M. Jerrum. Approximate counting, uniform generation and rapidly mixing Markov chains. *Information and Computation*, 82:93-133, 1989.

[40] F. Spitzer. *Principles of Random Walk.* Van Nostrand, 1964.

[41] A. Telcs. Random walks on graphs, electric networks and fractals. *Probab. Th. Rel. Fields*, 82:435-449, 1989.

[42] A. Telcs. Spectra of Graphs and Fractal Dimension I. *Probab. Th. Rel. Fields*, 85: 489 - 497.

[43] P. Whittle. *Systems in Stochastic Equilibrium.* Wiley, 1986.

[44] O. Yaron. *Random Walks on Trees*. Technical Report, Hebrew University, Jerusalem, 1988.

ANALYSIS OF INFECTIOUS DISEASE DATA FROM A SAMPLE OF HOUSEHOLDS

Niels G. Becker
La Trobe University, Australia.

Summary

Observations are made on the individuals from a sample of households at two points in time. At the first time point it is determined who is susceptible to a certain infectious disease. At the second time point it is determined which of the susceptibles have been infected. A method for the analysis of such data is derived with the aid of martingale theory and suitable epidemic model assumptions. In contrast to a previously proposed analysis the present approach requires only a modest amount of manual calculation and is based on less restrictive assumptions. An application to influenza data is given.

1. Introduction Consider the spread of an infectious disease through a community of households. Suppose the disease is transmitted primarily by sufficiently close contacts between susceptible and infectious individuals, and that such contacts may take place between any pair of individuals.

1.1. Data on Sizes of Outbreaks in Households. Each of the individuals from a random sample of households is tested at time 0 and again at time τ. At time 0 it is determined who is susceptible to the infectious disease and at time τ it is determined which of the susceptibles has been infected since time 0. The time interval $(0, \tau)$ might be the time interval during which the epidemic passes through this community, or the 'epidemic season' of a disease which is endemic in the community. It is important to consider this type of infectious disease data, because it is relatively easy to obtain and is often reliable. When based on tests of sera one tends to obtain more objective data and can hope to cope with the presence of subclinical infections.

It has been the practice to analyse such infectious disease data by assuming that outbreaks within households essentially evolve independently of each other. However, it seems safer to base the analysis on a model which permits both between and within household infection. Such an analysis is described by Becker & Hopper (1983) for a community in which every household is kept under observation. Complete observation is impractical for large communities. Haber et al. (1988) propose an analysis which can be applied to data from a random sample of households from a larger community. Here we demonstrate that the approach of Becker and Hopper applies under the assumptions used by Haber et al. There results an analysis which involves only simple calculations which are easily performed manually. Furthermore, the assumptions are not quite as restrictive as those of the model used by Haber et al.

The proposed analysis is derived using martingale theory for counting processes. For a review of the requisite results from that theory see Andersen & Borgan (1985). We are able to give explicit expressions for estimates and standard errors. Central limit theorems also exist, see Andersen & Borgan (1985), which permits hypothesis testing and the construction of confidence intervals.

1.2. Model Assumptions and the Basic Martingales. Consider a random sample of n households. Label these 1,2,...,n. Let $S_h(t)$ denote the number of individuals from household h who are still susceptible at time t. Write s_h for $S_h(0)$. Of the $N_h(t)=s_h-S_h(t)$ individuals infected by time t there are $I_h(t)$ infectious individuals at time t.

The infection intensity acting on each of the $S_h(t)$ susceptibles of household h at time t consists of two components. One arises from within the household and is given by $\beta I_h(t)$, while the other arises from the remaining community and is denoted by λ_t. Note that λ_t depends on the total number of infectives in the community at time t. This makes it more realistic to view λ_t as a random process. Randomness in λ_t can be accommodated within the present framework, but is excluded under the assumptions used by Haber et al. (1988).

Let F_t denote the history of the infection process up to time t. Then

$$M_{1h}(t) = N_h(t) - \beta\int_0^t I_h(x) S_h(x)\,dx - \int_0^t \lambda_x S_h(x)\,dx \quad h=1,2,\ldots,n$$

are orthogonal zero-mean martingales (ZMMs) with respect to the filtration $\{F_t\}$.

Further characteristics of the disease need to be specified. No specific assumptions are made about the latent period. We assume that the infectious period has a random duration and thereafter the infected individual becomes immune from further infection for the remaining duration of the observation period. Individuals who end their infectious period are called removals. Let $R_h(t)$ be the number of removals in household h at time t. In order to simplify the derivation we assume that the duration of the infectious period is exponential with rate parameter γ. This assumption is not crucial to the derivation and when this assumption is relaxed, see Becker (1979, 1981), one arrives at the same estimates. However, expressions for standard errors need to be modified slightly. The processes specified by

$$M_{2h}(t) = R_h(t) - \gamma\int_0^t I_h(x)\,dx \qquad h=1,2,\ldots,n$$

are orthogonal ZMMs, with respect to $\{F_t\}$, under the present assumptions.

The processes $M_{1h}(t)$ and $M_{2h}(t)$ h=1,2,...,n are the basic martingales used to derive the method of inference.

We will be integrating step functions with respect to ZMMs on several occasions. The left continuous modification of these step functions will be used on each occasion, without this being indicated explicitly by the notation. This is done to ensure that the integration is over a predictable process.

2. Community Acquired Infection. The risk of being infected from outside the household over the interval $(0, \tau)$ is determined by $\Lambda_\tau = \int_0^\tau \lambda_x dx$, the integrated global infection intensity. Consider the estimation of this quantity.

2.1. The Integrated Global Intensity Function. For each h define a counting process K_h by

$$K_h(t) = 1\,\{N_h(t) > 0\},$$

which simply indicates whether household h has been infected by time t. Note that $K_h(t)$ can be expressed as $\int_0^t 1\{N_h(x) = 0\}\,dN_h(x)$. We use this observation to construct the ZMMs

$$M_{1h}^{\dagger}(t) = \int_0^t 1\{N_h(x) = 0\}\,dM_{1h}(x)$$

$$= K_h(t) - \int_0^t \lambda_x s_h\{1 - K_h(x)\}\,dx \qquad h=1,2,\ldots,n.$$

The choice of $1\{N_h(x)=0\}$ as integrand is made so as to eliminate β from the expression. Its use restricts attention to the household only as long as it contains no infected individuals, which ensures that $1\{N_h(x)=0\}\, I_h(x)=0$.

By summing the $M_{1h}^{\dagger}$ over all households one obtains the ZMM

$$\kappa(t) = K_t - \int_0^t \lambda_x \left\{ s - \sum_h s_h K_h(x) \right\} dx,$$

where $s = \sum_h s_h$ and $K_t = \sum_h K_h(t)$, the total number of households infected by time t. A further integration leads to the ZMM

$$\kappa^*(t) = \int_0^t \frac{1\{K_x < n\}}{s - \sum_h s_h K_h(x)} d\kappa(x)$$

$$= \int_0^t \frac{dK_x}{s - \sum_h s_h K_h(x)} - \int_0^t \lambda_x 1\{K_x < n\} dx.$$

Equating this ZMM, evaluated at τ, to its mean indicates that

$$\hat{\Lambda}_\tau = \int_0^\tau \frac{dK_x}{s - \sum_h s_h K_h(x)} \tag{2.1}$$

is unbiased for $\int_0^\tau \lambda_x 1\{K_x < n\} dx$. The latter quantity is $\Lambda_\tau = \int_0^\tau \lambda_x dx$ unless every household contains at least one infected individual at time τ.

The bias $E\left(\hat{\Lambda}_\tau - \Lambda_\tau\right)$ is

$$E\int_0^\tau \lambda_x 1\{K_x = n\} dx = \int_0^\tau \lambda_x P(K_x = n) dx$$

$$= \int_0^\tau \lambda_x \prod_{h=1}^n (1 - e^{-s_h \Lambda_x}) dx,$$

which is a rapidly decreasing function of n. The bias is negligible in applications, where one should aim to have at least 50 households.

For the purpose of providing a standard error we need to consider the variance of κ*(t). From standard results in martingale theory, see Andersen & Borgan (1985, equations 3.15, 3.18), we find

$$V[\kappa^*(t)] = E\int_0^t \left[\frac{dK_x}{\left\{ s - \sum_h s_h K_h(x) \right\}^2} \right]. \tag{2.2}$$

Households of Equal Size. Suppose all households are of the same size, that is $s_h = m$ $h=1,2,\ldots,n$. Then

$$\hat{\Lambda}_\tau = \frac{1}{m}\int_0^\tau \frac{dK_x}{n-K_x} = \frac{1}{m}\sum_{j=0}^{K_\tau - 1} \frac{1}{n-j}$$

$$\cong \frac{1}{m} ln \left(\frac{n+\frac{1}{2}}{n-K_\tau+\frac{1}{2}} \right) \tag{2.3}$$

Also

$$\int_0^\tau \frac{dK_x}{\left\{ s - \sum_h s_h K_h(x) \right\}^2} = \frac{1}{m^2}\int_0^\tau \frac{dK_x}{(n-K_x)^2} = \frac{1}{m^2}\sum_{j=0}^{K_\tau - 1} (n-j)^{-2},$$

so that, in view of (2.2), we obtain

$$\text{s.e.}\,(\hat{\Lambda}_\tau) \cong \frac{1}{m}\left[\frac{1}{n-K_\tau+\frac{1}{2}} - \frac{1}{n+\frac{1}{2}} \right]^{\frac{1}{2}}. \tag{2.4}$$

Households of Varying Sizes. Often data are available from households of varying sizes. Strictly speaking, one is not able to compute $\hat{\Lambda}_\tau$ as given by (2.1), because the order in which the households are infected is not known. One way of overcoming this problem is to obtain a separate estimate of Λ_τ from each group of households of a given size. These estimates can then be pooled by taking their weighted average, with weights proportional to the inverse of their estimated variances. This should work well when the numbers in each group is moderately large. With small numbers there is a risk that a poor estimate receives a large weight.

An alternative approach is based on the observation that the value of (2.1) is much the same for all likely orderings of household infections. A good approximation to (2.1) is provided by calculating

$$m_a = \text{(mean size of affected households)}$$

and replacing $s - \sum_h s_h K_h(x)$ by $s - \sum_h m_a K_h(x)$. In other words, use

$$\hat{\Lambda}_\tau = \frac{1}{m_a}\int_0^\tau \frac{dK_x}{n^* - K_x} = \frac{1}{m_a}\sum_{j=1}^{K_\tau - 1}\frac{1}{n^* - j}$$

$$\cong \frac{1}{m_a} ln\left(\frac{n^* + \frac{1}{2}}{n^* - K_\tau + \frac{1}{2}}\right), \qquad (2.5)$$

where $n^* = s/m_a$.

By a similar argument

$$\text{s.e. } (\hat{\Lambda}_\tau) \cong \frac{1}{m_a}\left[\frac{1}{n^* - K_\tau + \frac{1}{2}} - \frac{1}{n^* + \frac{1}{2}}\right]^{1/2}. \qquad (2.6)$$

The expressions (2.5) and (2.6) agree exactly with (2.3) and (2.4), respectively, when all households are of the same size.

2.2. The Probability of Infection from Global Sources. The quantity $P_\tau = 1 - e^{-\Lambda_\tau}$ is of interest because of its meaningful epidemiological interpretation. It is the probability that an individual makes at least one infectious contact during $(0, \tau)$ with an infective from another household.

From data on outbreaks in n households of size m one can estimate P_τ by

$$\hat{P}_\tau = 1 - e^{-\hat{\Lambda}_\tau} \cong 1 - exp\left\{-\frac{1}{m} ln\left(\frac{n}{n - K_\tau}\right)\right\}$$

$$= 1 - \left(1 - \frac{K_\tau}{n}\right)^{1/m}.$$

The approximation holds when n is large. Using the delta method, see Rao (1973, section 6a.2), one obtains

$$\text{s.e. } (\hat{P}_\tau) = e^{-\hat{\Lambda}_\tau}\text{s.e. } (\hat{\Lambda}_\tau)$$

$$\cong e^{-\hat{\Lambda}_\tau}\frac{1}{m}\left[\frac{n}{n-K_\tau}-\frac{1}{n}\right]^{\frac{1}{2}}$$

$$=\frac{1}{m\sqrt{n}}\left(1-\frac{K_\tau}{n}\right)^{\frac{1}{m}-\frac{1}{2}}\left(\frac{K_\tau}{n}\right)^{\frac{1}{2}}.$$

These final expressions for $\hat{P}_\tau$ and s.e. $(\hat{P}_\tau)$ can also be arrived at by starting from the observation that, under the present model formulation, K_τ has a binomial distribution with index n and success probability parameter $e^{-m\Lambda_\tau}$. It is encouraging that the martingale approach leads to essentially the same estimates as the classical approach.

For households of varying sizes one simply replaces m and n by m_a and n*,respectively.

The estimate of Λ_τ given by this approach is essentially unbiased. By Jens

en's inequality it follows that $1-e^{-\hat{\Lambda}_\tau}$ will tend to underestimate $P_\tau = 1-e^{-\Lambda_\tau}$. However, it shares with the maximum likelihood estimate the property that it is asymptotically unbiased.

3. The Within - Households Infection Potential. There is no continuous observation over time and so one is not able to estimate the rate parameters β and γ. We turn instead to the estimation of the infection potential θ=β/γ. The interpretation of θ as an infection potential stems from the observation that it is of the form

(infection rate) x (mean duration of the infectious period).

Note that households of size 1 contain no information about θ. The discussion given below is therefore intended only for households having at least two susceptibles initially.

Consider again the basic martingales $M_{1h}(t)$ h=1,2,...,n. Let

$$H_{1h}(t) = 1\{K_t < n\}\ 1\{S_h(t) > 0\}/S_h(t)$$

and

$$H_{2h}(t) = 1\{K_t < n\}\ 1\{N_h(t) = 0\}\ (n - L_t)/\{s - \sum_h s_h K_h(t)\},$$

where L_t is the number of households with every individual infected by time t, so that $\sum_h 1\{S_h(t) > 0\} = n - L_t$.

Now construct the two ZMMs

$$M_1^*(t) = \sum_h \int_0^t H_{1h}(x)\, dM_{1h}(x)$$

$$= \sum_h \int_0^t \frac{1\{K_x < n\}}{S_h(x)} dN_h(x)$$

$$-\beta \sum_h \int_0^t 1\{K_x < n\}\, 1\{S_h(x) > 0\}\, I_h(x)\, dx$$

$$-\int_0^t 1\{K_x < n\}\, \lambda_x (n - L_x)\, dx$$

and

$$M_1^{\wp}(t) = \sum_h \int_0^t H_{2h}(x)\, dM_{1h}(x)$$

$$= \int_0^t \frac{n - L_x}{s - \sum_h s_h K_h(x)} dK_x - \int_0^t 1\{K_x < n\}\, \lambda_x (n - L_x)\, dx.$$

Note that the choices of H_{1h} and H_{2h} have made the terms involving λ_x the same in these two ZMMs. The quantity λ_x is therefore eliminated by taking the difference $M_1^*(t) - M_1^{\wp}(t)$. However, there is one other term which needs attention. The term containing β contains the unobservable $I_h(x)$ quantities. To eliminate these unobservable terms we use, for the first time, the ZMMs $M_{2h}(t)$ h=1,2,...,n.

Construct the ZMM

$$M_2^*(t) = \sum_h \int_0^t \theta 1\{K_x < n\}\, 1\{S_h(x) > 0\}\, dM_{2h}(x).$$

We finally obtain the ZMM

$$M_1^*(t) - M_1^{\wp}(t) - M_2^*(t)$$

$$= \sum_h \int_0^t \frac{1\{K_x < n\}}{S_h(x)} dN_h(x) - \int_0^t \frac{n - L_x}{s - \sum_h s_h K_h(x)} dK_x$$

$$-\theta \sum_h \int_0^t 1\{K_x < n\} 1\{S_h(x) > 0\} dR_h(x).$$

$$= A_t - B_t - \theta C_t, \text{ say.}$$

By evaluating this ZMM at τ and equating it to its mean we obtain the estimate

$$\hat{\theta} = (A_\tau - B_\tau)/C_\tau \quad .$$

It is not always possible to assign exact values to A_τ, B_τ and C_τ, but there are situations when one can.

Suppose $K_\tau < n$, which means that not every household becomes infected. Then

$$A_\tau = \sum_{h=1}^{n} \sum_{j=0}^{N_h(\tau)-1} \frac{1}{s_h - j}.$$

For B_τ and C_τ we substitute

$$n\hat{\Lambda}_\tau \cong \frac{n}{m_a} ln\left[\frac{n^* + 1/2}{(n^* - K_\tau + 1/2)}\right] \text{ and } R_\tau - 2L_\tau = \sum_{h=1}^{n} R_h(\tau) - 2L_\tau,$$

respectively. These are the exact values for B_τ and C_τ whenever no household is totally infected, that is when $L_\tau = 0$. Otherwise these values are likely to be larger than they should be and their use will tend to underestimate θ. For infectious diseases of low or moderate infectiousness L_τ will tend to be small, compared with n, and these values for B_τ and C_τ will be good approximations. The correction term $2L_\tau$ used for C_τ is somewhat arbitrary. For a household in which everyone gets infected, the term $\int_0^\tau 1\{K_x < n\} 1\{S_h(x) > 0\} dR_h(x)$ is essentially the number of removals at the time when the last susceptible is infected. This unobserved quantity takes values in $\{0,1,\ldots,R_h(\tau)-1\}$. However, its value can be at most $R_h(\tau)-2$ when the last infection arises from a within-household contact. We have, somewhat arbitrarily, proposed the value $R_h(\tau)-2$.

We now obtain an expression for the standard error of $\hat{\theta}$. Using the orthogonality of the ZMMs M_{1h} and M_{2h} h=1,2,...,n, we find

$$V[M_1^*(t)-M_1^{\wp}(t)-M_2^*(t)]$$

$$=V[M_1^*(t)]+V[M_1^{\wp}(t)]+V[M_2^*(t)]$$

$$-2\sum_h Cov\left[\int_0^t H_{1h}(x)\,dM_{1h}(x),\ \int_0^t H_{2h}(x)\,dM_{1h}(x)\right].$$

Now use known results from martingale theory, see Andersen & Borgan (1985, equations 3.15, 3.17 and 3.18), to find

$$V[M_1^*(t)-M_1^{\wp}(t)-M_2^*(t)]$$

$$=E\left[\sum_h\int_0^t H_{1h}^2(x)\,dN_h(x)\right]+E\left[\sum_h\int_0^t H_{2h}^2(x)\,dN_h(x)\right]$$

$$+\theta^2E\left[\sum_h\int_0^t 1\{K_x<n\}\,1\{S_h(x)>0\}\,dR_h(x)\right]$$

$$-2E\left[\sum_h\int_0^t H_{1h}(x)\,H_{2h}(x)\,dN_h(x)\right].$$

We need to estimate each of the four terms on the right hand side, with time t=τ. Unbiased estimates of these terms are given respectively, by

$$\sum_h\int_0^\tau\frac{1\{K_x<n\}}{S_h^2(x)}dN_h(x)\cong\sum_{h=1}^n\sum_{j=0}^{N_h(\tau)-1}\frac{1}{(s_{h-j})^2},$$

$$\int_0^\tau\frac{(n-L_x)^2}{\left\{s-\sum_h s_hK_h(x)\right\}^2}dK_x\cong n^2[s.\,e.\,(\hat\Lambda_\tau)]^2,$$

$$\hat\theta^2C_\tau\cong\hat\theta^2(R_\tau-2L_\tau),$$

$$2\sum_{h=1}^n\int_0^\tau\frac{(n-L_x)\,dK_h(x)}{s_h\left\{s-\sum_h s_hK_h(x)\right\}}\cong\frac{2}{m_a}B_\tau\cong\frac{2n}{m_a}\hat\Lambda_\tau.$$

The approximations given for each of the first three estimates are exact when at least one household escapes infection and no households are completely infected. However, they will be good approximations as long as the disease is not highly infectious, that is as long as L_τ is small compared with n. The same is true for the fourth estimate when all households are of the same size. The approximations for the fourth estimate rely on the additional approximation of setting $s_h = m_a$, for affected households, when the households are of varying sizes.

With these estimates we may compute

$$\text{s.e.}(\hat{\theta}) = \left\{ \hat{V}[M_1^*(\tau) - M_1^{\wp}(\tau) - M_2^*(\tau)] \right\}^{1/2} / C_\tau.$$

Suppose the disease is such that the infectious period has the same (constant) duration for each infective. In this case $\pi = 1 - e^{-\theta}$ is the probability that a given susceptible makes at least one infectious contact with a given infective from the same household during the latter's infectious period. Haber et al. (1988) refer to π as the secondary attack rate. In the case of a constant infectious period one estimates this by

$$\hat{\pi} = 1 - e^{-\hat{\theta}} \text{ and s.e.}(\hat{\pi}) = e^{-\hat{\theta}} \text{ s.e.}(\hat{\theta}).$$

An important difference is that, with a constant infectious period, one needs to set $V[M_2^*(\tau)] = 0$ in the computation of s.e. $(\hat{\theta})$.

4. Application to Influenza Data. Data on the sizes of outbreaks in Tecumseh households over the 1977-78 influenza epidemic season are given in Table 1. These are taken from Table 1 of Haber et al. (1988).

4.1. Community Acquired Infection. There are $K_\tau = 76$ affected households. The average size of affected households is $m_a = 206/K_\tau = 2.7105$. Substituting these values and $n^* = 616/m_a = 227.26$ into (2.5) and (2.6) gives

$$\hat{\Lambda}_\tau = 0.150 \text{ and s.e.}(\hat{\Lambda}_\tau) = 0.017.$$

This leads to

$$\hat{P}_\tau = 1 - e^{-\hat{\Lambda}_\tau} = 0.139 \text{ and s.e.}(\hat{P}_\tau) = e^{-\hat{\Lambda}_\tau} \text{ s.e.}(\hat{\Lambda}_\tau) = 0.015,$$

which compares exceptionally well with the maximum likelihood estimation 0.140 ± 0.015 given by Haber et al. (1988, Table 1).

It is instructive to compute an estimate of Λ_τ based only on the data from households of size j, for each j=1,2,3 and 4. The four estimates of Λ_τ obtained in this way are shown in column 4 of Table 2, with the corresponding standard errors in column 5. These estimates should be inspected for large differences and trends with increasing household size. Such anomalies would cast doubt on the underlying assumptions. No patterns or large differences are apparent in the current data set (unless there is a good reason why households of size two seem less prone to community acquired infection). The weighted average of the estimates of Λ_τ given in Table 2, with weights inversely proportional to the estimated variances, is 0.124 ± 0.014.

4.2. The Within-Household Infection Potential. We now concern ourselves only with households of size two, three and four. For these households we have n=200, s=538, $K_\tau=63$, $L_\tau=9$ and $m_a=193/K_\tau$. These lead to

$$A_\tau = 41.5,$$

$$B_\tau \cong \frac{n}{m_a} ln\left(\frac{n^* + 1/2}{n^* - K_\tau + 1/2}\right) = 28.9034$$

and $C_\tau \cong R_\tau - 2L_\tau = 79,$

from which we find

$$\hat{\theta} = (A_\tau - B_\tau)/C_\tau = 0.159.$$

The corresponding standard error is computed to be

$$\text{s.e.}\ (\hat{\theta}) = [21.875 + 13.4845 + 2.0085 - 18.8696]^{1/2}/79 = 0.054\ .$$

From these we obtain

$$\hat{\pi} = 1 - e^{-\hat{\theta}} = 0.147 \text{ and s.e. } (\hat{\pi}) = 0.044\ .$$

These compare reasonably well with the maximum likelihood estimate 0.155 ± 0.035 given by Haber et al. (1988, Table 1). Note that in the calculation of s.e.($\hat{\pi}$) we deleted the number 2.0085 from the above numerical expression for s.e.($\hat{\theta}$). This is appropriate when it is assumed that there is no variation in the duration of the infectious period.

When the within-household infection potential is estimated separately for households of size 2, 3 and 4 one obtains the estimates shown in Table 3. These should be inspected for large differences and trends, which provides a rough check of model assumptions. Although the estimates of θ decrease as m increases it is unwise to read too much into this because the standard errors are relatively large . The weighted average of these estimates of θ is also given in Table 3. This is seen to be close to the pooled estimate given above.

Acknowledgement. The author gratefully acknowledges support from the Danish Research Academy and the Statistical Research Unit, University of Copenhagen.

References

[1] Andersen, P.K. & Borgan, ϕ. (1985). Counting process models for life history data: a review. *Scand. J. Statist.* **12**, 97-158.

[2] Becker, N.G. (1979). An estimation procedure for household disease data. *Biometrika* **66**, 271-277.

[3] Becker, N.G.(1981). The infectiousness of a disease within households. *Biometrika* **68**, 133-141.

[4] Becker, N.G. & Hopper, J.L. (1983). The infectiousness of a disease in a community of households. *Biometrika* **70**, 29-39.

[5] Haber, M., Longini, I.M. & Cotsonis, G.A. (1988). Models for the statistical analysis of infectious disease data. *Biometrics* **44**, 163-173.

[6] Rao, C.R. (1973). Linear Statistical Inference and it Applications (2nd edition). New York: Wiley.

Table 1. Observed frequencies of outbreak sizes in Tecumseh during the 1977-1978 influenza epidemic season.

Number Infected	Initial number of susceptibles per household: 1	2	3	4
0	65	88	27	22
1	13	14	15	9
2		4	4	9
3			4	3
4				1
Total	78	106	50	44

Table 2. Requisite data and estimates for community acquired infection.

Household size	Frequency	Affected households K_τ	$\hat{\Lambda}_\tau$	s.e. $(\hat{\Lambda}_\tau)$
1	78	13	0.181	0.050
2	106	18	0.093	0.022
3	50	23	0.203	0.043
4	44	22	0.170	0.037
Total	278	76		

Table 3. Estimates, and associated standard errors, of the within-household infection potential for households of size 2, 3 and 4.

Household size, m	Infection potential $\hat{\theta}$	Standard error s.e. $(\hat{\theta})$
2	0.228	0.156
3	0.205	0.102
4	0.140	0.056
Weighted average	0.162	0.047

ON DIRECTED POLYMERS IN A RANDOM ENVIRONMENT

Erwin Bolthausen
Technische Universitat Berlin

We consider random walks whose laws are perturbed in an irregular way by a second random mechanism, the so-called random environment. The perturbation acts in such a way that visits to certain randomly chosen points are favoured or disfavoured. There have been proposed several models in the mathematical physics literature, sharing the common feature that only very few facts are known on a rigorous mathematical level. We present here some of these models and some problems connected with them.

We always look at a random walk on the d-dimensional lattice $\mathbf{Z}^d$ but we start by introducing the random environment, the "disorder". It is given by i.i.d. random variables X(i), $i \in \mathbf{Z}^d$, satisfying

$$X(i) > 0 \text{ almost surely} \tag{1}$$

$$EX(i) = 1. \tag{2}$$

X = 1 is then just the case where no perturbation occurs. Sometimes, it is convenient to have a one-dimensional parameter $\beta \in \boldsymbol{R}$ regulating the amount of disorder. This can be done by considering

$$X_\beta(i) = e^{\beta Y(i)}/m(\beta)$$

where Y(i) are i.i.d. real random variables such that $m(\beta) = E(e^{\beta Y}) < \infty$ for β in a neighborhood of 0.

The unperturbed random walk is an ordinary random walk $\xi_0 = 0, \xi_1, \ldots, \xi_T$ on $\mathbf{Z}^d$ whose jump distribution is given by p(x), $x \in Z^d$, $\Sigma_x p(x) = 1$, i.e. we have

$$P(\xi_1 = i_1, \ldots, \xi_T = i_T) = 1_0(\xi_0) \prod_{j=1}^{T} p(i_j - i_{j-1})$$

where $i_0 = 0$. We always assume that for some $\varepsilon > 0$, we have

$$\sum_{x \in Z^d} e^{\varepsilon|x|} p(x) < \infty \quad \text{and} \quad \Sigma\, xp(x) = 0.$$

This random walk and the random environment are chosen to be independent.

Formally, we consider the probability space $\Omega = \Omega_{\text{path}} \times \Omega_{env}$, $\Omega_{\text{path}} = (\mathbf{Z}^d)^{\mathbf{N}}$, $\Omega_{env} = (\mathbf{R})^{\mathbf{Z}^d}$, with the appropriate product field and equipped with the probability measure $Q = P \otimes \mu^{Z^d}$, where P is the law of the above defined random walk on $\mathbf{Z}^d$ and μ is the law of the X(i). The ξ and the X variables are given by the appropriate projections.

We denote by < > the expectation with respect to ξ and by E(.) the expectation with respect to the environment variables. Strictly speaking, < > is the conditional expectation given the X variables and E(.) is the conditional expectation given the random walk. The total expectation is then given by E(< >).

One natural model for a perturbation would be to introduce the weight factor

$$\kappa_T(X, \xi) = \prod_{j=1}^{T} X(\xi_j)$$

and to transform the measure P on the path space Ω_{path} for each realization of the X variables, by considering $\hat{P}_{T,X}$ defined by

$$d\hat{P}_{T,X}/dP(\omega) = \kappa_T(X,\xi(\omega))/\langle\kappa_T(X,\xi)\rangle, \qquad \omega \in \Omega_{\text{path}}$$

A typical quantity one is interested in is the mean square displacement

$$\int |\xi_T|^2 d\hat{P}_{T,X} = \langle |\xi_T^2| \kappa_T(X,\xi)\rangle / \langle k_T(X,\xi)\rangle$$

and its asymptotic behaviour for large T.

For the model just introduced, essentially nothing seems to be known. It is related to some problems for random Schrödinger operators. To see the relation, it is convenient to switch to a continuous time random walk which, for simplicity is assumed to be just a nearest neigbour symmetric random walk. Furthermore, we assume that X(i) is of the form exp(Y(i))/m, where the Y(i) are i.i.d. Bernoulli random variables taking values 1 or -1. Then

$$\kappa_T(X, \xi) = \exp\left(\int_0^T Y(\xi_s)\, ds\right)/m^T .$$

Assuming furthermore d=1, we consider the discrete Laplacian Δ on $\mathbf{Z}$: $\Delta f(i) = (f(i+1) + f(i-1) - 2f(i))/2$. It is known that for almost all realizations of the Y variables the operator $\Delta + Y$ acting on $l_2(\mathbf{Z})$ (Y just by multiplication) has a pure point spectrum which is dense in [-3,1] and where each eigenvalue has exponentially decaying eigenfunctions (see [2]). Therefore $\langle\kappa_T(X,\xi)1_i(\xi_T)\rangle$

can be expanded in terms of these eigenfunctions ψ_j:

$$< \kappa_T(X,\xi)\, 1_i(\xi_T) > \;=\; \sum_j e^{\lambda_j T}\; \psi_j(0)\, \psi_j(i)\,.$$

In the limit $T \to \infty$, only those eigenvalues count which are near the upper edge of the spectrum. However, there seems to be no information available, where the eigenfunctions are situated. Up to my knowledge, even for d = 1, the mean square displacement in the above model is not known.

The model of directed polymers in a random environment seems to have grown out of an even more difficult problem, where one replaces the random walk ξ by a polymer, i.e. a self avoiding random walk. There seems to exist absolutely no results for this on a mathematical level, but there is a considerable interest in this in the physics literature. Form background information, see the introduction of [3] and the references there.

A directed polymer is a caricature of a true polymer, where one makes the random walk self avoiding just by stepping in one of the dimensions deterministically one step in each time point (we take time discrete again). Of course, we can identify this special dimension with the discrete time axis. The model is therefore the same as that introduced in the beginning, with the only difference that the environment variables now change also independently in time, i.e. we consider random variables X(i,t), $i \in \mathbf{Z}^d$, $t \in \mathbf{N}$, which satisfy (1) and (2), and we put now

$$\kappa_T(X,\xi) \;=\; \prod_{s=1}^{T} X(\xi_s, s) \quad .$$

There is some disagreement what the "dimension" of such a model is. If the time axis is considered as one of the space dimensions, the dimension is d + 1.

It appears clear that this model is much easier than a true polymer in a random environment, but despite its simplicity, not much is known on a rigorous level. There seems to be the general belief that for d = 1 the mean square displacement of ξ_T is for almost all X realizations of order $T^{4/3}$ (see e.g. [5], [7]). The following result is a generalization of the results in [1] and [6]:

Theorem. *Let* d $\geq$ 3. *There exist* $\delta > 0$, *depending only on* d *and the transitions probabilities* p(x), *such that if* var(X) $\leq \delta$, *one has almost surely*

a) $$\lim_{T\to\infty} T^{-1} \sum_{i \in \mathbf{Z}^d} (\lambda, i)^2\, \mu_{T,X}(i) \;=\; (\Sigma\lambda,\lambda)$$

for all $\lambda \in \mathbf{R}^d$. *Here,* $(.,.)$ *denotes the inner product and* Σ *is the covariance matrix for p.*

b) $\mu_{T,X}(\sqrt{T}\,.) \to N(0,\Sigma)$ *in distribution*

This has been proved in [1] for the special case where $X = 1 + \varepsilon Z$, $Z = \pm 1$ with probability 1/2 and for the nearest neighbor random walk.

We prove here a) in our slightly more general situation. b) can be proved by showing the convergence of all moments.

To prove a), one first remarks that for any realization of the random walk ξ $\kappa_T(X,\xi)$ is an F_T- martingale, where $F_T = \sigma(X(i,s) : i \in \mathbf{Z}^d, s \le T)$, which satisfies $E(\kappa_T(X,\xi)) = 1$. It converges to 0 almost surely, except in the trivial case $X = 1$. Then also $<\kappa_T(X,\xi)>$ is an F_T-martingale, with $E(<\kappa_T(X,\xi)>) = 1$. The transience of the d-dimensional random walk for $d \ge 3$ is now used to prove that this martingale converges to a strictly positive limit if var(X) is small enough. To see this, we write

$$(E(\ <\kappa_T(X,\xi)>2) = E(<\kappa_T(X,\xi)\quad)><\kappa_T(X,\xi')>)$$

where ξ' is an independent copy of ξ. Interchanging E() with < >, we get

$$E(<\kappa_T(X,\xi)>^2) = \ <\prod_{s=1}^{T} E(X(\xi_s,s)X(\xi'_s,s))> \ = <(1+\mathrm{var}(X))^{n_T(\xi,\xi')}>$$

where

$$n_T(\xi,\xi') = \sum_{S=1}^{T} 1_{\xi_s=\xi'_s} = \sum_{S=1}^{T} 1_{\xi_s-\xi'_s=0} \le \sum_{S=1}^{\infty} 1_{\xi_s-\xi'_s=0} = n_\infty(\xi,\xi')$$

n_∞ obviously has an exponential moment, so, if var(X) is small enough, we have $\sup_T E(<\kappa_T(X,\xi)>^2) < \infty$ and, by the martingale convergence theorem, $<\kappa_T(X,\xi)>$ converges to a limit, say θ with $E(\theta) = 1$. The event

$$\{\lim_{T\to\infty} <\kappa_T(X,\xi)> \ = 0\}$$

is clearly measurable with respect to the field $\sigma(X(i,s) : i \in \mathbf{Z}^d, s \ge t)$ for any t, so, by the Kolmogorov 0-1-law, we get $P(\theta > 0) = 1$, showing that

$$P(\lim_{T\to\infty} <\kappa_T(X,\xi)> \text{ exists and is} > 0) = 1. \qquad (3)$$

To prove a) in our theorem, we use the well known fact that for any $\lambda \in \mathbf{R}^d$ $(\lambda,\xi_T)^2 - T(\lambda,\Sigma\lambda)$ is a martingale for the filtration of the random walk. Using that $\kappa_T(X,\xi)$ is an F_T-martingale for a fixed path, we conclude that

$$M_T = <((\lambda,\xi_T)^2 - T(\lambda,\Sigma\lambda))\kappa_T(X,\xi)>$$

is an F_T-martingale, and so

$$U_T = \sum_{s=1}^{T} \frac{1}{s}(M_s - M_{s-1}) \qquad (M_o = 0),$$

too. Remark that

$$M_s - M_{s-1} = <((\lambda,\xi_s)^2 - s(\lambda,\Sigma\lambda))\kappa_{s-1}(X,\xi)(X(\xi_s,s) - 1)>. \tag{4}$$

By a calculation which is slightly more complicated than that above for $E<\kappa_T(X,\xi)>^2$, using (4), one sees that if var(X) is small enough, one has

$$\sup_T E(U_T^2) < \infty,$$

so U_T converges almost surely, and by the Kronecker lemma, we have

$$\lim_{T\to\infty} \frac{1}{T} M_T = 0.$$

From this and (3), part a) in the theorem follows.

It is generally believed that the properties in the theorem no longer hold true if var(X) is too large, although this is essentially based only on Monte Carlo simulations (see [4] and the references cited there). This is related to a large deviation problem. We will give a short discussion of this point. To discuss this, it is convenient to have the one dimensional parameter β as introduced in the beginning, i.e. we consider

$$X_\beta(i,t) = e^{\beta Y(i,t)}/m(\beta)$$

where $m(\beta) = E(e^{\beta Y}) < \infty$ for $|\beta|$ small enough. Our theorem states that for small enough $|\beta|$, one has the properties a) and b). It is expected that even if Y is bounded (so $m(\beta)$ exists for all β) this central limit behavior breaks down for large enough $|\beta|$. This is not at all understood. The question is connected with large deviation problem in the following way: We have

$$\kappa_T(X,\xi) = \exp(\beta S_T)/m(\beta)^T,$$

where $S_T = \sum_{s=1}^{T} Y(\xi_s, s)$. From the above proved fact that for small enough $|\beta|$ one has

$$\lim_{T\to\infty} \langle \kappa_T(X,\xi) \rangle > 0$$

one concludes that

$$\lim_{T\to\infty} \frac{1}{T} \log \langle e^{\beta S_T} \rangle = \log m(\beta), \text{ almost surely.} \tag{5}$$

So, one has the somewhat strange fact that for almost all realizations of the environment variables Y, S_T has as a random variable depending on the path space the same large deviation behaviour as do sums of i.i.d. Y variables, at least in a neighborhood of EY. A first step towards an understanding of the large $|\beta|$ region would be a discussion of the question if (5) breaks down for large $|\beta|$. As far as I know, such a breakdown has not been proved. A strong indication that this happens is however given in a discussion of even simplified (so called mean-field) model by Derrida & Spohn [3]. They discussed a random walk on the so called Caley-tree which can be interpreted as a walk on an infinite dimensional lattice. The path space Ω_{path} here is the set of sequences

$$\xi_0 = 0, \xi_1, \xi_2, \ldots$$

with $\xi_t \in \{-1,1\}^t$, such that if $\xi_t = (a_1, \ldots, a_t)$ then $\xi_{t-1} = (a_1, \ldots, a_{t-1})$. The path measure is defined by the property that ξ_t adds to ξ_{t-1} as the t'th component +1 or -1 with probability 1/2.

The environment variables are given by

$$X_\beta(i) = \exp(\beta Y(i)) \text{ for } i \in A = \bigcup_{t=1}^{\infty} \{-1,1\}^t.$$

Then, one investigates

$$\kappa_T(X,\xi) = \prod_{s=1}^{T} X_\beta(\xi_s).$$

(It is unnecessary to take X(i,t) because the position in A already fixes the time point). Our analysis of $\langle\kappa_T(X,\xi)\rangle$ works as well here, so (5) is true for small enough $|\beta|$. Derrida and Spohn discussed the special case where Y is standard normally distributed, in which case $\log m(\beta) = \beta^2/2$. Here, in fact, there is a critical point for β and $\lim_{T\to\infty}(1/T) \log\langle e^{\beta S_T}\rangle$ agrees with $\beta^2/2$ only in some region $|\beta| \le \beta_{crit}$, and outside, this limit depends linearly on β.

If $|\xi_t|$ denotes the sum of the components in ξ_t then our analysis in the theorem works as well for this, proving e.g. that for small enough $|\beta|$

$$\lim_{T\to\infty} \frac{1}{T} \langle|\xi_T|^2\kappa_T(X,\xi)\rangle/\langle\kappa_T(X,\xi)\rangle = 1$$

for almost all realization of the environment variables. It would be interesting to know what happens with this quantity for $|\beta| > \beta_{crit}$.

References

[1] E. Bolthausen: A note on the diffusion of directed polymers in a random environment. Comm. Math. Phys. 123, 529-534 (1989).

[2] R. Carmona, A. Klein and F. Marinelli: Anderson localization for one dimensional random Schrödinger operators with Bernoulli and other singular potentials. Comm. Math. Phys. 108, 41 (1987).

[3] J. Cook and B. Derrida: Polymers on disordered hierarchical lattices: A nonlinear combination of random variables. To appear in "*J. Stat. Phys*".

[4] B. Derrida and H. Spohn: Polymers on disordered trees, spin glasses and travelling waves. J. Stat. Phys. 52, 609 (1988).

[5] D. Fisher, D. A. Huse and C. C. Henley: Exact exponents for an interface in a two dimensional random exchange Ising model Phys. Rev. Lett. 55, 2094 (1985).

[6] J. Z. Imbrie and T. Spencer: Diffusion of directed polymers in a random environment. J. Stat. Phys. 52, 609 (1988).

[7] J. Z. Imbrie: Directed plymers in a random environment. Math. quantum field theory and related topics. Can. Math. Soc. Conf. Proc. 9, 83-90 (1988).

A LOOK AT PERTURBATION APPROXIMATIONS FOR EPIDEMICS

H.E. Daniels
University of Cambridge

Abstract

Perturbation-type approximations have been derived by many authors for epidemics and related problems. Here we use the standard perturbation technique familiar to physicists to approximate the mean and variance of paths of some epidemic processes. For epidemics with no removals the saddlepoint approximation to the distribution of infectives is highly accurate for quite small populations and can be used to assess the accuracy of perturbation approximations. For epidemics with removals no saddlepoint approximation is available and simulations have to be used for comparison. The technique turns out to be equivalent to Daley and Kendall's method of 'diffusion of arbitrary constants'.

1. Introduction. Approximations of the perturbation type have been around in epidemic theory for a considerable time. Bartlett (1956) used such an approximation to study the fluctuations about the endemic equilibrium in a recurrent epidemic. Bailey (1968) and Weiss (1971) obtained by different routes essentially the same approximations for the mean and variance of a simple epidemic in large populations. Barbour (1972) developed the method of 'diffusion of arbitrary constants' introduced by Daley and Kendall (1965) to extend these results to more general epidemics. McNeil and Schach (1973) considered diffusion approximations of the Ornstein-Uhlenbeck type for epidemic and similar processes. Daniels (1960) described what is essentially a perturbation technique for approximating to the distribution and moments of processes of the epidemic type. Other references may be found in Bailey (1972) and Renshaw (1986).

There is, however, a standard procedure familiar to physicists and engineers for deriving perturbation approximations which can with advantage be used in problems of this kind. My interest in the method was first aroused by Bellman's elegant little monograph (Bellman, 1964) though he treats there only deterministic problems. In the present paper the procedure is applied systematically to obtain perturbation approximations for the mean and variance of some epidemic processes.

2. The Univariate Case. We first consider the univariate birth process which includes the so-called simple epidemic with no removals and its generalizations. At time t there are $N(t)$ infectives and the probability of a new infection in time $(t,t+dt)$ is $\lambda(N)dt$, ignoring terms which are $o(dt)$. If m is the total population size the number of susceptibles in the population is $m-N(t)$, and $\lambda(m)=0$. On a suitable time scale the simple epidemic has $\lambda(N)=N(m-N)$.

The process evolves according to the equation (Bartlett, 1960)

$$dN = \lambda(N)\ dt + dZ(t) \tag{2.1}$$

where $dZ(t)$ is a centred Bernoulli variable such that

$$P\{dZ(t) = 1-\lambda(N)\ dt\} = \lambda(N)\ dt$$

$$P\{dZ(t) = -\lambda(N)\ dt\} = 1-\lambda(N)\ dt$$

Initially $N(0) = a$. Then

$$E\{dZ(t)|N(t) = n\} = 0,\quad E\{[dZ(t)]^r|N(t)= n\} = \lambda(n)dt \text{ when } r \geq 2,$$

and

$$E\{dZ(t_1)\,dZ(t_2)|N(t_1) = n_1, N(t_2) = n_2\} = 0,\quad t_1 \neq t_2.$$

Since the moments of $dZ(t)$ involve n implicitly it is convenient to write

$$dZ(t) = \sqrt{\lambda(n)}\,dW(t)$$

in which case $E(dW(t)|n) = 0$, $E[(dW(t))^2|n] = dt$, but $E[(dW(t))^2|n] = dt/[\lambda(n)]^{\frac{1}{2}r-1}$, $r > 2$ still involves n. Since we are concerned only with the mean and variance of $N(t)$ we shall use a diffusion approximation where $W(t)$ is regarded as Brownian motion whose higher moments are free of n.

The basic idea underlying perturbation expansions is to embed (2.1) in the family of models

$$\begin{aligned} dN &= \lambda(N)\,dt + \varepsilon dZ(t) \\ &= \lambda(N)\,dt + \varepsilon\sqrt{\lambda(N)}\,dW(t)\,. \end{aligned} \tag{2.2}$$

When $\varepsilon = 0$ the model is deterministic with a known solution; when $\varepsilon = 1$ we have the stochastic model (2.1). It is then assumed that the solution of (2.2) can be formally expanded as a power series in ε

$$N = N_0 + \varepsilon N_1 + \varepsilon^2 N_2 + \ldots \tag{2.3}$$

which is substituted in (2.2). Thus

$$d(N_0 + \varepsilon N_1 + \varepsilon^2 N_2 + \ldots) = \lambda(N_0 + \varepsilon N_1 + \varepsilon^2 N_2 + \ldots)\,dt$$

$$+\varepsilon\sqrt{\lambda(N_0+\varepsilon N_1+\varepsilon^2 N+\dots)}\,dW(t)$$

$$=\lambda(N_0)\,dt+\varepsilon\lambda'(N_0)N_1dt\;+\varepsilon^2\left\{\lambda'(N_0)N_2+\frac{1}{2}\lambda''(N_0)N_1^2\right\}dt$$

$$+\{\varepsilon\sqrt{\lambda(N_0)}+\frac{1}{2}\varepsilon^2\frac{\lambda'(N_0)}{\sqrt{\lambda(N_0)}}N_1+\dots\;\}\,dW(t)$$

Equating coefficients of powers of ε we obtain the following equations

$$dN_0 = \lambda(N_0)\,dt$$

$$dN_1 = \lambda'(N_0)N_1dt+\sqrt{\lambda(N_0)}\,dW(t)$$

$$dN_2 = \lambda'(N_0)N_2dt+\frac{1}{2}\lambda''(N_0)N_1^2dt$$

$$+\frac{1}{2}\frac{\lambda'(N_0)}{\sqrt{\lambda(N_0)}}N_1dW(t)\,.$$

The first is the deterministic equation. Once its solution is found the others can be solved recursively.

The success of the procedure does not depend on ε being small. All that is needed is that the series (2.3) is a convergent or an asymptotic expansion for some range of ε which include $\varepsilon = 1$. Bellman refers to ε as a 'book keeping' variable allowing terms of comparable degrees of approximation to be marshalled in an orderly way. However, in the epidemic application the magnitude of successive terms is more clearly exhibited in terms of proportions $R = N/m$ rather than N. Initially $R = a/m = \alpha$. For simplicity we shall assume that $\lambda(N)$ has the form

$$\lambda(N) = m^c\beta(R)\,. \tag{2.4}$$

This includes the simple epidemic with $\lambda(N) = m^2R(1 - R)$ and Severo's more general form $\lambda(N) = N^j(m - N)^k = m^{j+k}R^j(1 - R)^k$ (Severo, 1969). Let $T = m^{c-1}t$, $dB(T) = m^{(c-1)/2}dW(t)$ and $\varepsilon = \omega\sqrt{m}$. Then (2.2) becomes

$$dR(T) = \beta(R)\,dT+\omega\sqrt{\beta(R)}\,dB(T) \tag{2.5}$$

where $EdB(T)=0$ $E\{dB(T)\}^2=dT$. The original stochastic epidemic is recovered on putting

$\omega=1/\sqrt{m}$ which is now small.

As before, write $R = R_0 + \omega R_1 + \omega^2 R_2 + \ldots$ and equate coefficients of powers of w, obtaining

$$dR_0 = \beta(R_0)\,dT$$

$$dR_1 = \beta'(R_0)\,R_1 dT + \sqrt{\beta(R_0)}\,dB(T)$$

$$dR_2 = \beta'(R_0)\,R_2 dT + \frac{1}{2}\beta''(R_0)\,R_1^2 dT + \frac{1}{2}\frac{\beta'(R_0)}{\sqrt{\beta(R_0)}}R_1 dB(T)\,.$$

The first equation has the deterministic solution

$$T = \int_\alpha^{r_0} \frac{dx}{\beta(x)} \tag{2.6}$$

where r_0 has been written in place of R_0 since it is deterministic. The second equation can then be solved using the integrating factor

$$e^{-\int \beta'[r_0(T)]\,dT} = e^{-\int \frac{\beta'(r_0)}{\beta(r_0)}dr_0} = \frac{1}{\beta(r_0)}$$

which yields

$$R_1(T) = \beta[r_0(T)]\int_0^T \frac{dB(u)}{\sqrt{\beta[r_0(u)]}}\,. \tag{2.7}$$

It follows that $E\{R_1(T)\} = 0$ and

$$E\{R_1^2(T)\} = \beta^2[r_0(T)]\int_0^T \frac{du}{\beta[r_0(u)]}$$

or, in terms of r_0,

$$E(R_1^2) = \beta^2(r_0)\int_\alpha^{r_0} \frac{dx}{\beta^2(x)}\,. \tag{2.8}$$

Using the same integrating factor the third equation yields

$$R_2(T) = \frac{1}{2}\beta[r_0(T)]\int_0^T \beta''[r_0(u)]\,R_1^2(u)\,du$$

$$+\frac{1}{2}\beta[r_0(T)]\int_0^T \frac{\beta'[r_0(u)]\,R_1(u)\,dB(u)}{\beta^{\frac{3}{2}}[r_0(u)]} \tag{2.9}$$

Notice that in the second term $dB(u)$ is an innovation and should strictly be written $dB(u+)$. Hence $E\{R_1(u)dB(u)\} = 0$ and the second term vanishes on taking expectations. After some manipulation it is found that

$$E(R_2) = \frac{1}{2}\beta(r_o)\beta'(r_0)\int_\alpha^{r_0}\frac{dx}{\beta^2(x)} + \frac{1}{2}\left\{1-\frac{\beta(r_0)}{\beta(\alpha)}\right\} \tag{2.10}$$

which can be written as

$$E(R_2) = \frac{1}{2}\frac{\beta'(r_0)}{\beta(r_0)}E(R_1^2) + \frac{1}{2}\left\{1-\frac{\beta(r_0)}{\beta(\alpha)}\right\}. \tag{2.11}$$

Finally, on inserting $\omega = 1/\sqrt{m}$ we obtain

$$E(R) = r_0 + \frac{1}{m}E(R_2) + 0\left(m^{-\frac{3}{2}}\right) \tag{2.12}$$

$$\text{var}R = \frac{1}{m}E(R_1^2) + 0\left(m^{-\frac{3}{2}}\right) \tag{2.13}$$

where $E(R_1^2)$ and $E(R_2)$ are given by (2.8) and (2.11). It follows from the general theory of birth processes that, necessarily, $E(R_2) < 0$.

A useful consequence of the fact that $E\{R_1(u)\,dB(u)\} = 0$ in (2.9) is that as far as terms in ω^2 we could equally well have taken $E\{(dZ)^2 | n\}$ to be $\lambda(n_0)\,dt$, thus avoiding the need to expand $\sqrt{\lambda(N)}$ in the last term of (2.2). This fact is used later when discussing the general epidemic with removals.

In terms of the actual number of susceptibles we have

$$n_0 = mr_0, \quad N_1 = \sqrt{m}R_1, \quad N_2 = R_2 \tag{2.14}$$

and the corresponding formulae are

$$t = \int_\alpha^{n_0}\frac{dx}{\lambda(x)}$$

$$E(N_1^2) = \lambda^2(n_0)\int_\alpha^{n_0}\frac{dx}{\lambda^2(x)} \tag{2.15}$$

$$E(N_2) = \frac{1}{2}\frac{\lambda'(n_0)}{\lambda(n_0)}E(N_1^2) + \frac{1}{2}\left\{1 - \frac{\lambda(n_0)}{\lambda(\alpha)}\right\}.$$

The approximations are then

$$\begin{aligned} E(N) &\sim n_0 + E(N_2) + 0\left(m^{-\frac{1}{2}}\right) \\ \mathrm{var}N &\sim E(N_1^2) + 0\left(m^{-\frac{1}{2}}\right) \end{aligned} \qquad (2.16)$$

3. Two Examples. We now calculate the perturbation approximations to the mean and variance for two particular models. The first is the simple epidemic with $\lambda(n) = n(m-n)$ where the results are found to agree with those obtained by other methods. The second is that suggested by McNeil (1972) with $\lambda(n) = n^{1/2}(m-n)^{1/2}$ as a simplified model for a spatial epidemic.

Except in special cases, or for very small populations, the exact results are difficult to calculate for comparison. However, it was shown (Daniels, 1982) that the saddlepoint approximation to the distribution of the number of infectives, even for quite small population sizes, is almost indistinguishable from the exact distribution. The corresponding means and variances have therefore been used for comparison. The paper by Daniels (1982) may be consulted for details of the saddlepoint method.

The simple epidemic $\lambda(n) = n(m-n)$

The computations are most easily carried out in terms of R, with $\beta(r) = r(1-r)$. The deterministic solution is $r_0/(1-r_0) = \{\alpha/(1-\alpha)\}e^t$ and it is found from (2.8) that

$$E(R_1^2) = r_0^2(1-r_0)^2\left\{2T - \frac{1}{r_0} + \frac{1}{1-r_0} + \frac{1}{\alpha} - \frac{1}{1-\alpha}\right\},$$

from which $E(R_2)$ and $E(R)$ can be calculated using (2.11) and (2.12). The formulae agree with those given by Daniels (1960, p.398) and Bailey (1968).

The corresponding results for $E(N)$ and $SD(N) = \sqrt{\mathrm{var}N}$ are compared with the exact values (based on the saddlepoint approximations) in Table 1 for m=20 and a=1,5,10. The deterministic values n_0 are also shown. Notice first that when a=1, excluding the initial stages when the process behaves like a linear birth process, the perturbation approximation to the mean and SD goes seriously wrong. This is a well known feature of diffusion approximations, which cannot properly account for very small initial values. When a=5 and a=10 the values are in good agreement. There is a tendency for the approximate values of $SD(N)$ to drop towards the completion of the epidemic but as the distribution is then very skew the SD is perhaps not very appropriate.

The McNeil model $\lambda(n) = n^{1/2}(m-n)^{1/2}$.

A simplified spatial epidemic is modelled by considering the infections to take place on the circumference of an expanding circle within which all are infected. In terms of $R, \beta(r) = r^{1/2}(1-r)^{1/2}$ and the deterministic solution is

$$r_0 = \sin^2(\sin^{-1}\sqrt{\alpha} + \tfrac{1}{2}T).$$

With this model the epidemic is completed in a finite time $T(\max) = \pi - 2\sin^{-1}\sqrt{\alpha}$. From (2.8)

$$E(R_1^2) = r_0(1-r_0)\log\left\{\frac{r_0(1-r_0)}{\alpha(1-\alpha)}\right\}$$

$$E(R_2) = \frac{1}{4}\frac{(1-2r_0)}{r_0(1-r_0)}ER_1^2 + \frac{1}{2}\left\{1-\left[\frac{r_0(1-r_0)}{\alpha(1-\alpha)}\right]^{\frac{1}{2}}\right\}$$

from which $E(R)$ and var$R$, and hence $E(N)$ and $SD(N)$ are calculated as before.

Comparisons with the exact mean and SD, and the deterministic value n_0 are shown in Table 2 for m=20, a=1 and 5. In this case even the results for a=1 are quite good.

4. The Bivariate Case A similar technique is next applied to the bivariate process

$$\left.\begin{aligned} dX &= \phi(X, Y)\,dt + \varepsilon dG(t) \\ dY &= \psi(X, Y)\,dt + \varepsilon dH(t) \end{aligned}\right\} \qquad (4.1)$$

which becomes the required process when $\varepsilon = 1$. We shall not specify $dG(t)$, $dH(t)$ in detail at the moment. In the next section the results are applied to the so-called general epidemic with removal of infectives where the appropriate $dG(t)$ and $dH(t)$ are defined.

As before, the solution of (4.1) is formally expanded as

$X = X_0 + \varepsilon X_1 + \varepsilon^2 X_2 \tau \ldots$, $Y = Y_0 + \varepsilon Y_1 + \varepsilon^2 Y_2 + \ldots$ and the coefficients of powers of ε are equated. From ε^0 we obtain

$$\frac{dX_0}{dt} = \phi(X_0, Y_0), \quad \frac{dY_0}{dt} = \psi(X_0, Y_0). \tag{4.2}$$

It is assumed (which is often not the case) that (4.2) has a known deterministic solution $x_0(\xi, \eta, t)$, $y_0(\xi, \eta, t)$ where ξ, η are the initial values of X_0, Y_0, i.e.

$$x_0(\xi, \eta, 0) = \xi, \qquad y_0(\xi, \eta, 0) = \eta \quad .$$

The terms in ε^1 are

$$\left.\begin{aligned} dX_1 &= \phi'(x_0, y_0) X_1 dt + \dot{\phi}(x_0, y_0) Y_1 dt + dG(t) \\ dY_1 &= \psi'(x_0, y_0) X_1 dt + \dot{\psi}(x_0, y_0) Y_1 dt + dH(t) \end{aligned}\right\} \tag{4.3}$$

when $\phi' = \dfrac{\partial\phi(x, y)}{\partial x}$, $\dot{\phi} = \dfrac{\partial\phi(x, y)}{\partial y}$, and so on. In the univariate case the equivalent equation was solved by using an integrating factor. Here we use the method of variation of parameters.

The homogeneous equations

$$\left.\begin{aligned} \frac{dX_1}{dt} &= \phi'(x_0, y_0) X_1 + \dot{\phi}(x_0, y_0) Y_1 \\ \frac{dY_1}{dt} &= \psi'(x_0, y_0) X_1 + \dot{\psi}(x_0, y_0) Y_1 \end{aligned}\right\} \tag{4.4}$$

are the linearised form of (4.3). They have two independent solutions

$$\left(\frac{\partial x_0}{\partial \xi}, \frac{\partial y_0}{\partial \xi}\right) \text{ and } \left(\frac{\partial x_0}{\partial \eta}, \frac{\partial y_0}{\partial \eta}\right),$$

as may be discovered on substituting $x_0(\xi + \delta\xi, \eta + \delta\eta, t)$, $y_0(\xi + \delta\xi, \eta + \delta\eta, t))$ in (4.2), expanding and extracting the coefficients of $\delta\xi$ and $\delta\eta$.

The complete solution of (4.4) is therefore

$$\left(a_1\frac{\partial x_0}{\partial \xi} + a_2\frac{\partial x_0}{\partial \eta}\right), \quad \left(a_1\frac{\partial y_0}{\partial \xi} + a_2\frac{\partial y_0}{\partial \eta}\right).$$

The solution of the nonhomogeneous equations (4.3) is then found by allowing a_1 and a_2 to be functions of t in

$$\left.\begin{aligned} X_1(t) &= a_1(t)\frac{\partial x_0}{\partial \xi} + a_2(t)\frac{\partial x_0}{\partial \eta} \\ Y_1(t) &= a_1(t)\frac{\partial y_0}{\partial \xi} + a_2(t)\frac{\partial y_0}{\partial \eta} \end{aligned}\right\}. \tag{4.5}$$

On substituting in (4.3) it is found that $a_1(t)$, $a_2(t)$ must satisfy

$$\left.\begin{aligned} \frac{\partial x_0}{\partial \xi}da_1(t) + \frac{\partial x_0}{\partial \eta}da_2(t) &= dG(t) \\ \frac{\partial y_0}{\partial \xi}da_1(t) + \frac{\partial y_0}{\partial \eta}da_2(t) &= dH(t) \end{aligned}\right\}. \tag{4.6}$$

The solution of (4.6) is

$$\begin{aligned} a_1(t) &= \int_0^t \frac{\left\{\dfrac{\partial y_0(u)}{\partial \eta}dG(u) - \dfrac{\partial x_0(u)}{\partial \eta}dH(u)\right\}}{\dfrac{\partial[x_0(u), y_0(u)]}{\partial(\xi, y)}} \\ a_2(t) &= \int_0^t \frac{\left\{-\dfrac{\partial y_0(u)}{\partial \xi}dG(u) + \dfrac{\partial y_0(u)}{\partial \xi}dH(u)\right\}}{\dfrac{\partial[x_0(u), y_0(u)]}{\partial(\xi, \eta)}}. \end{aligned} \tag{4.7}$$

Hence

$$\begin{aligned} X_1(t) &= \int_0^t \frac{\dfrac{\partial[x_0(t), y_0(u)]}{\partial(\xi, \eta)}dG(u) - \dfrac{\partial[x_0(t), x_0(u)]}{\partial(\xi, \eta)}dH(u)}{\dfrac{\partial[x_0(u), y_0(u)]}{\partial(\xi, \eta)}} \\ Y_1(t) &= \int_0^t \frac{-\dfrac{\partial[y_0(t), y_0(u)]}{\partial(\xi, \eta)}dG(u) + \dfrac{\partial[y_0(t), x_0(u)]}{\partial(\xi, \eta)}dH(u)}{\dfrac{\partial[x_0(u), y_0(u)]}{\partial(\xi, \eta)}} \end{aligned} \tag{4.8}$$

where all the Jacobians are known functions calculated from the deterministic solution.

The terms in ε^2 lead to the equations

$$
\begin{aligned}
dX_2 &= \phi'(x_0, y_0)\, X_2 dt + \dot{\phi}(x_0, y_0)\, Y_2 dt \\
&\quad + \left\{ \frac{1}{2}\phi''(x_0, y_0)\, X_1^2 + \dot{\phi}'(x_0, y_0)\, X_1 Y_1 + \frac{1}{2}\ddot{\phi}(X_0, y_0)\, Y_1^2 \right\} dt \\
dY_2 &= \psi'(x_0, y_0)\, X_2 dt + \dot{\psi}(x_0, y_0)\, Y_2 dt \\
&\quad + \left\{ \frac{1}{2}\psi''(x_0, y_0)\, X_1^2 + \dot{\psi}'(x_0, y_0)\, X_1 Y_1 + \frac{1}{2}\ddot{\psi}(x_0, y_0)\, Y_1^2 \right\} dt \ .
\end{aligned}
\tag{4.9}
$$

They are solved in precisely the same way, the solution for $X_2(t)$, $Y_2(t)$ being (4.8) with $dG(u) = \{\frac{1}{2}\phi''(x_0,y_0)X_1^2 + \dot{\phi}'(x_0,y_0)X_1Y_1 + \frac{1}{2}\ddot{\phi}(x_0,y_0)Y_1^2\}dt$ and $dH(u)$ the corresponding expression with ψ for ϕ. Then $E(X_1^2)$, $E(X_1Y_1)$, $E(Y_1^2)$ and $E(X_2)$, $E(Y_2)$ are found on taking expectations. Just as in the univariate case, as far as terms in ε^2 it can without loss be assumed that $E\{[dG(u)]^2 | X,Y\} = x_0 y_0 dt$ and so on.

We now apply these results to the general epidemic.

5. The General Epidemic With Removals. The population consists of $X(t)$ susceptibles and $Y(t)$ infectives. The infection rate is taken to be $X(t)Y(t)$ as in the simple epidemic, but infectives are removed at rate $\rho Y(t)$ during the course of the epidemic. The appropriate equations are

$$
\begin{aligned}
dX &= -XYdt - dZ_X \\
dY &= (X-\rho)\, Ydt + dZ_X - dZ_Y
\end{aligned}
\tag{5.1}
$$

where

$$
E(dZ_X | X,Y) = 0, \quad E(dZ_Y | X,Y) = 0, \quad E\{(dZ_X)^2 | X,Y\} = XYdt,
$$

$$
E\{(dZ_Y)^2 | X,Y\} = \rho Ydt, \quad E[dZ_X dZ_Y | X,Y] = 0.
$$

Initially $X(0) = \xi$, $Y(0) = \eta$. Here $\phi(X,Y) = -XY, \psi(X,Y) = (X-\rho)Y$ and $dG = -dZ_X$, $dH = dZ_X - dZ_Y$.

The deterministic solution is x_0, y_0 where

$$
\left.
\begin{aligned}
t &= \int_{x_0}^{\xi} \frac{dx}{x\{\rho \log(x/\xi) \;\; x \;\; \xi \;\; \eta\}} \\
y_0 &= \rho \log(x_0/\xi) - x_0 + \xi + \eta
\end{aligned}
\right\} .
\tag{5.2}
$$

It is convenient to rearrange (4.8) in the form

$$\left.\begin{aligned} X_1(t) &= \int_0^t \{-A_x dZ_x(u) + B_X dZ_Y(u)\} \\ Y_1(t) &= \int_0^t \{A_Y dZ_X(u) - B_Y dZ_Y(u)\} \end{aligned}\right\} \tag{5.3}$$

where

$$\begin{aligned} A_X(t,u) &= \frac{\partial[x_0(t), x_0(u)+y_0(u)]}{\partial(\xi,\eta)} \Big/ \frac{\partial[x_0(u), y_0(u)]}{\partial(\xi,\eta)} \\ B_X(t,u) &= \frac{\partial[x_0(t), x_0(u)]}{\partial(\xi,\eta)} \Big/ \frac{\partial[x_0(u), y_0(u)]}{\partial(\xi,\eta)} \\ A_Y(t,u) &= \frac{\partial[y_0(t), x_0(u)+y_0(u)]}{\partial(\xi,\eta)} \Big/ \frac{\partial[x_0(u), y_0(u)]}{\partial(\xi,\eta)} \\ B_Y(t,u) &= \frac{\partial[y_0(t), x_0(u)]}{\partial(\xi,\eta)} \Big/ \frac{\partial[x_0(u), y_0(u)]}{\partial(\xi,\eta)}. \end{aligned} \tag{5.4}$$

Then

$$\begin{aligned} E(X_1^2) &= \int_0^t \{A_X^2 x_0(u) + B_X^2 \rho\} y_0(u)\,du \\ E(X_1 Y_1) &= \int_0^t \{A_X A_Y x_0(u) + B_X B_Y \rho\} y_0(u)\,du \\ E(Y_1^2) &= \int_0^t \{A_Y^2 x_0(u) + B_Y^2 \rho\} y_0(u)\,du \end{aligned} \tag{5.5}$$

The solution for X_2, Y_2 is obtained by taking

$$dG(u) = -X_1(u)Y_1(u)du, \qquad dH(u) = X_1(u)Y_1(u)du$$

in (4.8). Then

$$X_2 = -\int_0^t A_X X_1(u)\,Y_1(u)\,du, \quad Y_2 = \int_0^t A_Y X_1(u)\,Y_1(u)\,du \tag{5.6}$$

from which $E(X_2)$, $E(Y_2)$ can be found using $E(X_1Y_1)$ as described at the end of section 9. The expectations have to be evaluated by numerical integration.

From (5.2), $\frac{\partial x_0}{\partial \xi} = x_0 y_0 \{1/\xi\eta - (1-\rho/\xi)\,w(x_0)\}$, $\frac{\partial x_0}{\partial \eta} = -x_0 y_0 w(x_0)$, where

$$w(x_0) = \int_{x_0}^{\xi} \frac{dx}{x\{\rho \log(x/\xi) \quad x \quad \xi \quad \eta\}^2} \tag{5.7}$$

Also $\frac{\partial y_0}{\partial \xi} = (1-\rho/\xi) - (1-\rho/x_0)\frac{\partial x_0}{\partial \xi}, \frac{\partial y_0}{\partial \eta} = 1 - (1-\rho/x_0)\frac{\partial x_0}{\partial \eta}$. These can be used to evaluate the Jacobians. In particular the denominator of A_X etc. has the simple form $\partial[x_0(u), y_0(u)]/\partial(\xi, \eta) = x_0(u)y_0(u)/\xi\eta$. As in the univariate case it was also found easier to work in terms of x_0 rather than t in the integrations using $dx_0(u) = x_0(u)y_0(u)\,du$.

Calculations were carried out for $\xi = 100$, $\eta = 5$, $\rho = 25$. With these values the probability of an epidemic failing to develop initially is $(\rho/\xi)^\eta = 0.001$ which can be neglected. Unfortunately no saddlepoint approximation is available for the exact values, which have therefore had to be estimated by simulation. Table 3 compares these with the perturbation approximations for $E(X)$, $SD(X)$ and $E(Y)$, $SD(Y)$.

In practice, the number of infectives Y at any time in the population cannot be directly observed. What can be observed is the number of removals $Q = \xi + \eta - X - Y$. A table has therefore been included of $E(Q)$ and $SD(Q) = \{\text{var}X + \text{var}Y + 2\text{cov}XY\}^{1/2}$, and their perturbation approximations.

There is reasonably good agreement with the exact values for $E(X)$, $E(Y)$ and $E(Q)$, but as in the case of the simple epidemic the standard deviations do not match up so well.

6. General Comments. We have chosen to deal with the particularly simple case where a diffusion approximation suffices for the first and second moments. If higher moments are needed, for example, in the univariate case, powers of $m^{-1/2}$ appear in the expectations which are not accounted for by the powers of ω. The method can still be used but care is needed in deciding which powers of $m^{-1/2}$ have to be retained at each stage. Similar considerations arise when $\lambda(N)$ does not have the simple form (2.4).

This observation also applies to an approach like that of McNeil and Schach (1973) which is based on the characteristic function but still employs a diffusion approximation. On the other hand the methods used by Bailey (1968), Weiss (1971) and Daniels (1960) do not require diffusion approximations.

The application of the method to the bivariate equation in Section 4 will be recognised as essentially the same as the method of 'diffusion of arbitrary constants' introduced by Daley and Kendall (1965). Professor David Williams has also pointed out to me that the treatment of the linearised equations (4.3) is formally equivalent to the Malliavin calculus currently employed in the study of large deviations.

Acknowledgement

I am greatly indebted to my colleague Dr G.A. Young for performing the calculations leading to Table 3.

References

[1] Bailey, N.T.J. (1968). 'A perturbation approximation to the simple stochastic epidemic in a large population'. *Biometrika* **55**, 199-209.

[2] Bailey, N.T.J. (1975). *The Mathematical Theory of Infectious Diseases.* Griffin.

[3] Barbour, A.D. (1972). 'The principle of the diffusion of arbitrary constants'. *J. Appl. Prob* **9**, 519-541.

[4] Bartlett, M.S. (1956). 'Deterministic and stochastic models for recurrent epidemics'. *Proc. Third Berkeley Symp. Math. Statist. and Prob.* **4**, 81-109. Berkeley and Los Angeles. University of California Press.

[5] Bartlett, M.S. (1960). *Stochastic Population Models.* Chapman and Hall.

[6] Bellman, R. (1964). *Perturbation Techniques in Mathematics, Physics and Engineering.* Holt, Rinehart and Winston.

[7] Daley, D.J. and Kendall, D.G. (1965). 'Stochastic rumours'. J. *Inst. Math. Appl.* **1**, 42-55.

[8] Daniels, H.E. (1960). 'Approximate solutions of Green's type for univariate stochastic processes'. *J. Roy. Statist. Soc. (B)* **22**, 376-401.

[9] Daniels, H.E. (1982). 'The saddlepoint approximation for a general birth process'. *J. Appl. Prob* **19**, 20-28.

[10] McNeil, D.R. (1972). 'On the simple stochastic epidemic'. *Biometrika* **59**, 494-497.

[11] McNeil, D.R. and Schach, S. (1973). 'Central limit analogues for Markov population processes'. *J. Roy. Statist. Soc. (B)* **35**, 1-23.

[12] Renshaw, E. (1986). 'A survey of stepping-stone models in population dynamics'. *Adv. Appl. Prob.* **18**, 581-626.

[13] Severo, N.C. (1969). 'Generalizations of some stochastic epidemic models'. *Math. Biosci.* **4**, 395-402.

[14] Weiss, G.H. (1971). 'On a perturbation method in the theory of epidemics'. *J. Appl. Prob.* **19**, 20-28.

TABLE 1

Simple epidemic. $\lambda(n) = n(m - n), T = mt$

$m = 20, a = 1$

Deterministic		Exact		Perturbation	
T	n_0	$E(N)$	$SD(N)$	$E(N)$	$SD(N)$
0.2	1.21	1.21	0.49	1.21	0.50
0.5	1.61	1.59	0.93	1.59	0.96
1.0	2.50	2.44	1.71	2.43	1.84
2.0	5.60	5.05	3.42	4.90	4.11
3.0	10.28	8.61	4.74	8.06	5.59
4.0	14.84	12.30	5.08	11.96	4.65
5.0	17.73	15.33	4.53	15.75	2.72
6.0	19.10	17.42	3.53	18.13	1.39
7.0	19.66	18.77	2.54	19.25	0.71
8.0	19.87	19.25	2.17	19.87	0.39

$m = 20, a = 5$

Deterministic		Exact		Perturbation	
T	n_0	$E(N)$	$SD(N)$	$E(N)$	$SD(N)$
0.5	7.09	7.07	1.55	7.06	1.58
1.0	9.51	9.38	2.27	9.37	2.36
1.5	11.98	11.70	2.63	11.69	2.74
2.0	14.22	13.82	2.67	13.80	2.71
2.5	16.05	15.59	2.47	15.57	2.40
3.0	17.40	16.96	2.14	16.96	1.97
3.5	18.34	17.96	1.76	17.98	1.55
4.0	18.96	18.66	1.40	18.69	1.18
4.5	19.36	19.13	1.10	19.16	0.90

$m = 20, a = 10$

Deterministic		Exact		Perturbation	
T	n_0	$E(N)$	$SD(N)$	$E(N)$	$SD(N)$
0.1	10.50	10.50	0.71	10.50	0.71
0.2	11.00	10.99	0.98	10.99	0.99
0.4	11.97	11.96	1.35	11.95	1.37
0.6	12.91	12.88	1.58	12.87	1.61
0.8	13.80	13.74	1.72	13.73	1.76
1.0	14.62	14.53	1.80	14.53	1.83
1.5	16.35	16.21	1.78	16.21	1.80
2.0	17.62	17.46	1.59	17.46	1.57
2.5	18.48	18.33	1.35	18.33	1.30
3.0	19.05	18.93	1.10	18.93	1.03

Table 2

McNeil model. $\lambda(n) = \sqrt{n(m-n)}$, $T = t$

$m = 20, a = 1$

Deterministic		Exact		Perturbation	
T	n_0	$E(N)$	$SD(N)$	$E(N)$	$SD(N)$
0.25	2.36	2.31	1.33	2.30	1.39
0.50	4.19	4.02	2.20	3.99	2.32
0.75	6.39	6.04	2.94	6.01	3.08
1.00	8.81	8.26	3.51	8.25	3.65
1.25	11.30	10.56	3.85	10.56	3.97
1.50	13.71	12.77	3.94	12.80	4.01
1.75	15.87	14.77	3.75	14.83	3.75
2.00	17.71	16.45	3.33	16.52	3.18
2.25	19.05	17.73	2.77	17.71	2.32
2.50	19.82	18.64	2.18	18.23	1.17

$m = 20, a = 5$

Deterministic		Exact		Perturbation	
T	n_0	$E(N)$	$SD(N)$	$E(N)$	$SD(N)$
0.25	7.28	7.28	1.57	7.28	1.59
0.50	9.76	9.69	2.25	9.69	2.29
0.75	12.24	12.09	2.64	12.09	2.72
1.00	14.59	14.32	2.76	14.33	2.87
1.25	16.64	16.23	2.59	16.26	2.75
1.50	18.28	17.70	2.20	17.74	2.33
1.75	19.41	18.72	1.70	18.64	1.62
2.00	19.96	19.34	1.21	18.61	0.57

Table 3

General epidemic. $X(t)$ susceptibles, $Y(t)$ infectives
$X(0) = \xi = 100$, $Y(0) = \eta = 5$, Removal rate $\rho = 25$

Susceptibles

Deterministic		Simulation		Perturbation	
t	x_0	$E(X)$	$SD(X)$	$E(X)$	$SD(X)$
1	92.90	92.99	4.06	92.98	4.13
2	80.48	81.10	8.85	81.14	9.73
3	62.84	64.99	13.36	65.13	13.92
4	43.90	47.85	15.52	47.93	14.92
5	28.46	33.19	14.86	32.82	12.40
6	18.14	22.45	12.68	21.63	9.05
7	11.89	15.27	10.21	14.35	6.44
8	8.41	10.70	8.10	9.87	4.74
9	5.99	7.80	6.53	7.16	3.68
10	4.62	5.95	5.44	5.47	3.01
11	3.74	4.75	4.73	4.38	2.58
12	3.15	3.94	4.28	3.71	2.29
13	2.74	3.41	3.99	3.20	2.09
14	2.46	3.04	3.80	2.92	1.96

Table 3 continued

Infectives

	Deterministic	Simulation		Perturbation	
t	y_0	$E(X)$	$SD(X)$	$E(X)$	$SD(X)$
1	10.26	10.18	4.04	10.19	4.10
2	19.09	18.54	7.93	18.50	8.24
3	30.55	28.75	10.85	28.63	11.22
4	40.52	37.55	11.29	37.50	10.63
5	45.12	42.15	9.62	42.52	7.94
6	44.19	42.25	7.72	42.95	6.43
7	39.88	39.19	6.70	39.84	6.31
8	34.29	34.53	6.42	34.92	6.30
9	28.64	29.42	6.28	29.58	6.03
10	23.51	24.55	6.06	24.51	5.58
11	19.09	20.16	5.67	20.03	5.05
12	15.40	16.40	5.17	16.54	4.52
13	12.36	13.23	4.64	13.29	4.01
14	9.90	10.64	4.13	11.25	3.55

Table 3 continued

Removals $Q = \xi + \eta - X - Y$

	Deterministic	Simulation		Perturbation	
t	q_0	$E(X)$	$SD(X)$	$E(X)$	$SD(X)$
1	1.84	1.83	1.25	1.84	1.26
2	5.43	5.36	2.39	5.36	2.43
3	11.62	11.25	4.21	11.24	4.39
4	20.58	19.60	6.39	19.57	6.61
5	31.41	29.67	8.26	29.66	8.17
6	42.67	40.30	9.41	40.43	8.73
7	53.23	50.53	9.76	50.22	8.54
8	62.31	59.77	9.53	60.21	7.93
9	70.36	67.78	8.98	68.27	7.16
10	76.87	74.51	8.20	75.02	6.37
11	82.17	80.09	7.48	80.58	5.64
12	86.45	84.65	6.72	84.74	4.99
13	89.89	88.36	6.09	88.50	4.53
14	92.64	91.32	5.54	90.83	3.96

A CENTRAL LIMIT THEOREM FOR EVOLVING RANDOM FIELDS

P. E. Greenwood and M. Ossiander
University of British Columbia and Oregon State Universisty

Abstract

A functional central limit theorem is proved for certain random fields whose domain has both a temporal and a spacial component. These processes are made up of dependent summands which are measurable with respect to an increasing filtration. Temporal limit theory for semi-martingales is utilized to provide spacial finite-dimensional convergence. Consequently the limiting random fields have independent increments in time, and can be thought of as evolving random fields. In deriving a tightness result the notion of majorizing measures is employed to allow local spacial variability. Thus a functional central limit theorem for evolving random fields with a rather general dependence structure is given here for the first time. Comparison with results available for empirical processes suggest that this result is close to optimal.

Introduction. In this paper we consider conditions for weak convergence of a sequence of random fields which are individually evolving over time. For motivation though, let's first examine a given sequence of random fields at some fixed time point. Let **X** be an arbitrary index set and for each $n \geq 1$, let $\{Y_{n,i}(x) : 1 \leq i \leq n,\ x \in \mathcal{X}\}$ be a sequence of random functions $Y_{n,i} : \mathcal{X} \to \mathbb{R}$. Let

$$S_n(x) = \sum_{i \leq n} Y_{n,i}(x) \qquad \text{for} \qquad x \in \mathcal{X}.$$

Thus the sequence $\{S_n(x) : n \geq 1,\ x \in \mathcal{X}\}$ is a sequence of random fields. We are interested in general conditions under which S_n converges weakly. We wish to allow dependence among the individual summands, the $Y_{n,i}$'s. We do this by considering conditions of the martingale type, involving only conditional first and second moments. (These are more desirable than mixing type conditions which involve the entire distribution.) In addition, we must constrain our indexing family X to satisfy a complexity condition involving the notion of majorizing measures (or, alternatively, metric entropy). There will be a natural topology on **X** associated with our problem.

In two papers Goldie and Greenwood (1986a, b) found conditions for weak convergence of sequences of set-indexed random fields. An unexpectedly difficult aspect of the problem was the characterization of the limiting distribution in a way which allowed the identification of the limiting finite-dimensional distributions. The first of the two papers was devoted to this problem, and the solution there was rather unsatisfactory in that mixing conditions were used rather than martingale type conditions.

An entirely different way of characterizing finite-dimensional laws is available if one considers, instead of a sequence of random fields $\{S_n(x) : x \in X\}$, a sequence of *evolving* random fields

$$S_n(x, t) = \sum_{i \leq \tau(n, t)} Y_{n,i}(x) \quad \text{for} \quad x \in X,\ t \in [0, 1].$$

(Here $\tau(n,t)$ depends on both n and t, perhaps randomly.) The idea of an evolving random field is appealing for modeling because most naturally occurring random fields are, in fact, either evolving over time or have arrived at the state in which we observe them by development through time. There are many examples where an evolving random field model is appropriate and weak limit theory is needed. Picture empirical data being gathered on a d-dimensional screen through real time. This example has been studied extensively as a random field, with time fixed at the point when sampling ends, but is also of interest as an evolving random field. A limit result can now be stated which allows dependence in both space and time of the empirical data. Another kind of example is a measure-valued process obtained from a branching diffusion. Here a functional limit theorem for the evolving random field has not yet been obtained (using our topology) but the conditions to be checked can be seen using our Theorem 2.1. Further applications arise related to e.g. U-statistics, chi-squared tests for spatial processes, and robust estimation for autoregressive process (see Koul (1989)).

An evolving random field has a natural partial ordering along the time axis. Each finite collection of points $\underset{\sim}{x} = (x_1, \ldots, x_k) \subset X$ corresponds to a sequence of k-dimensional stochastic processes in the time parameter:

$$S_n(\underset{\sim}{x}, t) = (S_n(x_1, t), \ldots, S_n(x_k, t)),\ t \in [0, 1].$$

The proof of weak convergence of evolving random fields will reduce to proving *finite-dimensional convergence* (i.e. weak convergence of $\{S_n(\underset{\sim}{x}, t) : t \in [0, 1]\}$) and *tightness* of $\{S_n(x,t) : x \in X,\ t \in [0, 1]\}$. (See Theorem 2.1.)

If each $S_n(\underset{\sim}{x}, t)$ is a Markov process in t we can prove weak convergence using the theory found, e.g. in the book of Ethier and Kurtz (1986). For a variety of models, results about weak convergence of interacting particle systems as in the book of Liggett (1985) might be used. If $S_n(\underset{\sim}{x}, t)$ is a sequence of semimartingales with respect, e.g., to the filtration $F_{n,t} = \sigma\{S_n(x,s) \ : x \in X,\ s \leq t\ \}$ then conditions for convergence of the finite-dimensional processes $S_n(\underset{\sim}{x}, t)$ appear in the book of Jacod and Shiryaev (1987).

In this paper we find conditions for weak convergence of a sequence of evolving random fields where the finite-dimensional processes $\{S_n(\underset{\sim}{x}, t) \ : t \in [0,1]\ \}$, are semimartingales. The sense of weak convergence here is different from and, in fact, much stronger than the sense usually employed in the subject called "measure-valued processes" or "super-processes." There, the processes are considered as *indexed* by t, only, and *evaluated* on x. The topology relative to which convergence is proved is then much weaker than the one we consider and in fact, by a theorem of Mitoma (c.f. Walsh (1984)), weak convergence of $S_n(x,t)$, for fixed $x \in X$ is sufficient for tightness of the "super-processes." Here our main challenge will be to prove tightness. The weak convergence obtained is, of course, correspondingly more useful.

The requisite tightness, or asymptotic equicontinuity, component of the weak convergence of $\{S_n(x,t) : x \in X, t \in I\}$ is obtained by restricting the complexity of **X** with a *majorizing measure condition* in conjunction with a *bracketing condition.* The concept of majorizing measures dates back to the early 1970's, see Preston (1972) and Fernique (1974). Fernique (1974) gave a characterization of **X**, together with its canonical metric, in terms of majorizing measures which was sufficient for the a.s. continuity of a Gaussian process $\{Z(x) : x \in X\}$. Quite recently Talagrand (1986) gave a necessary condition for the a.s. continuity of Z in terms of majorizing measures. (See Theorem 1.1.) The use of a majorizing measure condition in this context allows more local variability in the richness of X than a metric entropy condition would allow.

Now let $\Delta^{(\delta)}_{n,i}(\ \cdot\ ) = \left\{ \sup_{y : \sigma(\cdot, y) < \delta} |Y_{n,i}(y) - Y_{n,i}(\ \cdot\ )| \right\}^*$ where, for $G \in B(X)$, G^* denotes the P-outer envelope of G. Conditions which restrict the magnitude of the $\Delta^{(\delta)}_{n,i}$'s are termed *bracketing conditions.* The use of bracketing conditions first arose in the study of the central limit theorem for the set-indexed empirical measure, under the name *inclusion conditions* for families of sets. Dudley (1978) was one of the first to use them in developing tightness results for empirical processes; tying the L_p variability of the brackets (for p > 2) to constraints on metric entropy. The development of a new method of proof of asymptotic equi-continuity allowed Ossiander (1985) and (1986) to impose only L_2 constraints on the variability of the brackets through the use of a simple metric entropy condition. Using a refinement of the same method, while relying upon the recent work of Talagrand (1987), this last result was further improved by Andersen, Giné, Ossiander and Zinn (AZOG) (1988) by linking constraints on the weak-L_2 variability of the brackets to (natural) majorizing measure conditions. Ledoux and Talagrand (1989) have a comprehensive study of recent results in this area. A key ingredient of all of the above work is the assumption of independence of the $Y_{n,i}$'s which is not made here. Instead the $Y_{n,i}$'s are taken to be measurable with respect to an increasing filtration and the following key assumptions on the $Y_{n,i}$'s are made:

(i) uniform *conditional* asymptotic negligibility,

(ii) uniform *conditional* asymptotic centering,

(iii) a *conditional* bound on the weak-L_2 norm, uniform over a collection of brackets.

Thus in this paper functional central limit theorems for evolving random fields with a rather general dependence structure are obtained for the first time.

In section 1 we introduce notation and discuss the continuity of Gaussian processes. Some relevant results involving majorizing measures are presented. A definition of weak convergence and conditions implying the weak convergence of evolving random fields are given in section 3. This section contains a statement of our main central limit theorem, as well as several variants. Finite dimensional convergence and some characteristics of the limiting Gaussian process are discussed in section 3. Section 4 contains some exponential probability bounds for martingales. The crucial tightness result is proved in section 5. It depends on an exponential probability bound which may be of independent interest.

1. Notation, Definitions, and Preliminaries. Let(Ω, M, P) be a complete probability space and let **X** be some (arbitrary) topological space. Let B(**X**) denote the set of all bounded real-valued functions on **X** and **B(X)** denote the smallest σ-field containing all sets of the form $\{f \in B(X) : f(x_j) \in B_j, j = 1, ..., m\}$ where m is an arbitrary integer, the x_j are points in **X**, and the B_j are half-open intervals in $\mathbb{R}$.

Definition 1.1. A *random field*, Y, on X is a measurable mapping Y from (Ω, M) into (B(X), **B**(X)). Thus Y(x,*w*) denotes the value of the function in B(X) picked out by *w* at the point x. Generally, we suppress *w* through-out and simply write Y(x).

We assume throughout that the random fields we work with are separable in the following sense.

Definition 1.2. A *separable* random field is a random field, Y, for which there exists a countable set $D \subset X$ and a fixed event N for which P(N) = 0, such that for any closed interval $C \subset R$ and open set $U \subset X$ the two sets $\{w: Y(x, w) \in C,\ x \in U\}$ and $\{w : Y(x, w) \in C,\ x \in U \cap D\}$ differ (at most) by a subset of N.

For each $n \geq 1$, let $\{F_{n,i} : i \geq 1\}$ be an increasing filtration on the probability space (Ω, F, P). Let $F_{n,0} = \{\emptyset, \Omega\}$ be the trivial σ-algebra. For each $n \geq 1$ and $t \in I = \{t : 0 \leq t \leq 1\}$ let τ(n,t) be a stopping time with respect to $\{F_{n,i} : i \geq 0\}$. Also let $\{Y_{n,i} : i \geq 1\}$ be an array of separable random fields on **X** with each $Y_{n,i}$ being measurable with respect to $F_{n,i}$. Let

$$S_n(x, t) = \sum_{1 \leq i \leq \tau(n, t)} Y_{n,i}(x) \text{ for } x \in X \text{ and } t \in I.$$

Note that S_n is a random field on $\mathbf{X} \times \mathbf{I}$.

The next definition is standard.

Definition 1.3. A (centered) *Gaussian* random field on **X** is a random field on **X** for which each finite linear combination $\sum_{1 \leq j \leq m} a_j Y(x_j)$ is a centered Gaussian random variable. That is, each linear combination has a density of the form $(2\pi\sigma^2)^{-1/2} \exp\{-y^2/2\sigma^2\}$ for some $\sigma > 0$.

We utilize the following two definitions in discussing our limiting process. Let $\mathbf{I} = \{t : 0 \leq t \leq 1\}$ denote the unit inteval, and let $\{F_t : 0 \leq t \leq 1\}$ be an increasing filtration contained in *F*. That is, $F_s \subset F_t \subset F$ for $0 \leq s < t \leq 1$.

Definition 1.4. A F_t*-evolving* random field, Z, on $\mathbf{X} \times \mathbf{I}$ is a random field on $\mathbf{X} \times \mathbf{I}$ for which $Z(\cdot,t)$ is measurable with respect to F_t for each $t \in \mathbf{I}$ and has independent increments in **I** uniformly over **X**; that is for any x, $y \in X$, Z(x,t) is independent of Z(y,s) - Z(y,t) for $0 \leq t < s \leq 1$.

Definition 1.5. A *standard* F_t-evolving random field, **Z**, on $X \times I$ is a F_t-evolving random field on $X \times I$ for which

$$EZ(x,t) = 0,$$

$$EZ^2(x,0) = 0.$$

and

$$E(Z(x,s) - Z(x,t))^2 = |s - t| \, EZ^2(x,1)$$

for all $x \in X$ and $s, t \in I$. We say that Z is a (*standard*) *evolving random field* if Z is a (standard) F_t-evolving random field for some increasing filtration $\{F_t : 0 \le t \le 1\}$ contained in F.

Clearly a Gaussian random field Z on $X \times I$ is a standard evolving random field if and only if $Z(x, \cdot)/(EZ^2(x,1))^{1/2}$ is a standard Brownian motion for each fixed x.

The following two definitions are due to Fernique (1974).

Definition 1.6. Let d be a (pseudo-) metric on **X**. A Borel (sub-) probability measure ν on **X** is a *majorizing measure* for **X** with respect to d if

$$\sup_{x \in X} \int_0^\infty ln^{1/2}\ (1/\nu(B_d(x,u)))\, du < \infty. \tag{1.1}$$

Here, $B_d(x,u) = \{y \in X : d(x,y) < u\}$ is the u-ball in (**X**,d) centered at x.

Definition 1.7. A sub-probability measure μ is a *discrete majorizing measure* if it is a majorizing measure with support being a countably dense subset of **X**.

Let $\{Z(x) : x \in X\}$ be a mean zero Gaussian process on the space **X**. Let d denote the canonical L_2 pseudo-metric generated on **X** by **Z**, so

$$d(x,y) = (E(Z(x) - Z(y))^2)^{1/2}. \tag{1.2}$$

The following theorem is due to Fernique (1974) and Talagrand (1987).

Theorem 1.1. Suppose that the Gaussian process $\{Z(x) : x \in X\}$ is separable in (**X**,d). Then: (i) (Fernique) If (**X** ,d) admits a majorizing measure, ν, then Z has bounded sample paths a.s.. If, in addition,

$$\lim_{\delta \to 0} \sup_{x \in X} \int_0^\delta ln^{1/2} (1/\nu\ (B_d(x,u))\ )\, du = 0 \tag{1.3}$$

then the sample paths of Z are uniformly continuous with respect to d a.s.

(ii) (Talagrand) If Z has bounded sample paths a.s., then (**X**,d) admits a majorizing measure. If the sample paths of Z are uniformly continuous with respect to d a.s. then (**X**,d) admits a majorizing measure ν, satisfying (1.3.).

It is easy to see that if (**X**,d) admits a majorizing measure ν, it also must admit a discrete majorizing measure μ. If (1.3) holds for ν, the following lemma, due to Andersen (1986), allows us to assume that μ has a particular structure.

Lemma 1.1 If (**X**,d) admits a majorizing measure for which (1.3) holds, then for any $\beta \in (0,1)$ there exists a discrete majorizing measure μ satisfying the following. For each $x \in X$ and $k \geq 0$,

$$\text{there exists } x_k \text{ in the support of } \mu \text{ with } d(x,x_k) < \beta^k, \tag{1.4}$$

and

$$\lim_{K\to\infty} \sup_{x\in X} \sum_{k\geq K} \beta^k ln^{1/2}(1/\mu(x_k)) = 0. \tag{1.5}$$

We may also assume that

$$\text{if } x_k = y_k \text{ for a pair } x, y \in X, \text{ then } x_j = y_j \text{ for } j < k. \tag{1.6}$$

Furthermore

$$\lim_{K\to\infty} \sup_{x\in X} \sum_{k\geq K} \beta^k ln^{1/2}(1/\beta^k \prod_{j\leq k}\mu(x_j)) = 0. \tag{1.7}$$

Notice that (1.5) is the discrete analogue of (1.3), and that (1.7) follows easily from (1.5) using Kronecker's Lemma and interchange of summation. The following relationship between majorizing measures and metric entropy is useful.

Definition 1.8. The metric entropy of **X** with respect to d is given by

$H(\delta,X,d) = \ln \min\{N :$ there exists $x_1, \ldots, x_n \in X$ such that $d(x_i,x) < \delta$ for every $x \in X$ for some $i = 1, \ldots, n\}$.

It is easy to verify the following lemma.

Lemma 1.2. If

$$\int_0^1 H^{1/2}(u, X, d)\, du < \infty \tag{1.8}$$

then there exists a majorizing measure μ such that (1.3) holds.

2. Weak Convergence and the Central Limit Theorem. In this section we see how weak convergence of $\{S_n(x,t) : x \in X,\ t \in I\}$ follows from finite-dimensional convergence and asymptotic equi-continuity (tightness). We use the definitions and viewpoint of Gaenssler and Schneemeier (1986). An alternate approach using the Hoffmann-Jorgensen definition of weak convergence (which would yield a slightly more general result) is that of Andersen and Dobrić (1987). Results of later sections are called in as needed in order to state our main results here.

Let $l^\infty(X,I) = \{f : X \times I \to R : \sup_{(x,t) \in X \times I} |f(x,t)| < \infty\}$ denote the space of real-valued bounded functions on $X \times I$ and equip it with the supremum norm, so that $\|f\|_{X \times I} = \sup_{(x,t) \in X \times I} |f(x,t)|$. For any metric r on $X \times I$, let $g_r = \{f \in l^\infty(X,I) : f$ is uniformly r-continuous on $X \times I\}$. For a mean zero standard evolving Gaussian process $\{Z(x,t) : x \in X, t \in I\}$, let σ be the canonical L_2 metric on $X \times I$ given by

$$\sigma((x,s),(y,t)) = (E(Z(x,s) - Z(y,t))^2)^{1/2}. \tag{2.1}$$

Let d be the metric on **X** given by

$$d(x,y) = \sigma((x,1)(y,1)) \tag{2.2}$$

and ρ be the metric on **I** given by

$$\rho(s,t) = |s-t|^{1/2} \sup_{x \in X} (\mathrm{Var}\ Z(x,1))^{1/2}. \tag{2.3}$$

Letting τ denote the metric on $X \times I$ given by

$$\tau((x,s),(y,t)) = d(x,y) + \rho(s,t), \tag{2.4}$$

it is easy to see that τ dominates σ. If Z is separable and uniformly $\sigma(\tau)$-continuous, then from Talagrand's result (Theorem 1.1 (ii)) we know that $(X \times I, \sigma)$ $((X \times I, \tau))$ is totally bounded. Then $g_\sigma (g_\tau)$ is separable and closed in $(l^\infty(X \times I), \|\cdot\|)$; c.f. Corollary 2 of Gaenssler and Schneemeier (1986). Let **B** be the σ-algebra of all Borel sets in $(l^\infty(X \times I), \|\cdot\|)$ and let $\mathbf{B}_b$ denote the sub-σ-algebra of **B** generated by the open $\|\cdot\|$-balls in $l^\infty(X \times I)$.

Let $\{S_n : n \geq 1\}$ and Z all be defined on a common probability space (Ω, F, P), with each S_n being an element of $(X \times I, \mathbf{B})$ and Z being an element of $(l^\infty(X \times I), \mathbf{B}_b)$.

Definition 2.1. The sequence S_n *converges in distribution* to Z ($S_n \overset{D}{\rightarrow} Z$) if and only if

$$P(Z \in g_\sigma) = 1 \tag{2.5}$$

and

$$\lim_{n \to \infty} Ef(S_n) = Ef(Z) \tag{2.6}$$

for all $f \in \{g : \mathbf{X} \times \mathbf{I} \to R : g$ is bounded, uniformly $\|\cdot\|$-continuous and $\mathbf{B}_b$-measurable$\}$.

Theorem 2.1. If $\{Z(x,t) : x \in \mathbf{X}, t \in \mathbf{I}\}$ is a standard Gaussian evolving random field with

$$(\mathbf{X},d) \text{ admitting a majorizing measure which satisfies (1.3),} \tag{2.7}$$

for any finite collection $x_1, ..., x_k \in \mathbf{X}$

$$(S_n(x_1, \cdot), ..., S_n(x_k, \cdot)) \overset{D}{\rightarrow} (Z(x_1, \cdot), ..., Z(x_k, \cdot)), \tag{2.8}$$

and, for all $\eta > 0$,

$$\lim_{\delta \to 0} \overline{\lim_{n}} P\left[\sup_{t \in I, d(x,y) < \delta} |S_n(x,t) - S_n(y,t)| > \eta\right] = 0 \tag{2.9}$$

then $S_n \overset{D}{\rightarrow} Z$.

Proof: Condition (2.8) of the theorem identifies the limiting distribution and using Lemma 3.2, (2.7) guarantees its continuity. In view of the domination of σ by τ, Theorem 2.12 of Andersen and Dobrić (1987), and Theorem 1 of Gaenssler and Schneemeier (1986), it remains for us to show that (2.9) (together with (2.6), (2.7), and (2.8)) imply that for any $\eta > 0$,

$$\lim_{\delta \to 0} \overline{\lim_{n \to \infty}} P^*\left[\sup_{1 \le i \le I_\delta, \tau((x_i,t_i),(x,t)) < \delta} |S_n(x_i,t_i) - S_n(x,t)| > \eta\right] = 0,$$

where $\{(x_i,t_i) : 1 \le i \le I_\delta\}$ is a (finite) δ-net in $(\mathbf{X},\tau)$. Note that

$$B_\tau((x,t),\delta) \subset B_d(x,\delta) \times B_\rho(t,\delta)$$

and

$$|S_n(x_i,t_i) - S_n(x,t)| \le |S_n(x_i,t_i) - S_n(x_i,t)| + |S_n(x_i,t) - S_n(x,t)|.$$

Thus

$$P^*[\sup_{1\le i\le I_\delta, \tau((x_i,t_i),(x,t))<\delta} |S_n(x_i,t_i) - S_n(x,t)| > \eta]$$

$$\le P^*[\sup_{1\le i\le I_\delta, d(x_i,x)<\delta, t\in \mathbf{I}} |S\ (x_i,t) - S_n(x,t)| > \eta/2]$$

$$+ P^*[\sup_{1\le i\le I_\delta, \rho(t_i,t)<\delta} |S_n(x_i,t_i) - S_n(x_i,t)| > \eta/2]$$

$$\le P^*[\sup_{t\in \mathbf{I}, d(x,y)<\delta} |S_n(x,t) - S_n(y,t)| > \eta/2]$$

$$+ P^*[\sup_{1\le i\le I_\delta, \rho(s,t)<\delta} |S_n(x_i,s) - S_n(x_i,t)| > \eta/2].$$

From (2.9), if we let n go to infinity, and then δ go to zero, the first term goes to zero. From (2.8) (together with (2.6)),

$$\lim_{n\to\infty} P^*\left[\sup_{1\le i\le \mathbf{I}_\delta, \rho(s,t)<\delta} |S_n(x_i,s) - S_n(x_i,t)| > \eta/2\right]$$

$$= P^*\left[\sup_{1\le i\le \mathbf{I}_\delta, \rho(s,t)<\delta} |Z(x_i,s) - Z(x_i,t)| > \eta/2\right].$$

Now from (2.5) we see that this term decreases to zero as δ decreases. □

Notice that a result parallel to Theorem 2.1 is easily available for many cases when the limiting distribution is that of a general evolving Gaussian random field. The needed ingredient is a metric τ, dominating the canonical metric σ, with τ, being of the form given in (2.4), where d is as defined in (2.2) and $\rho(s,t) = \sup_{x\in\mathbf{X}} (E(Z(x,s) - Z(x,t))^2)^{1/2}$ with Z(x,t) being uniformly ρ-continuous in t over **X**.

The following corollaries are easy consequences of Theorem 2.1 and Theorem 5.1, 5.2, and 5.3 respectively.

Corollary 2.1. If (2.7) holds,

(3.1), (3.2), and (3.3) of Lemma 3.1 (finite-dimensional convergence) are satisfied, (2.10)

and

conditions (5.7), (5.8), (5.9) and (5.10) of Theorem 5.1 giving the uniform modulus of continuity are satisfied, (2.11)

then $S_n \xrightarrow{D} Z$.

Note that the definition of convergence in distribution implies that the Gaussian process **X** is a.s. σ-continuous on $\mathbf{X} \times \mathbf{I}$. Other variants are:

Corollary 2.2. If (2.7) and (2.10) hold, as well as

(5.12) and (5.13) of Theorem 5.2, (2.12)

then $S_n \xrightarrow{D} Z$.

Corollary 2.3. If (2.7) and (2.10) hold, as well as

(5.12) and (5.14) of Theorem 5.3, (2.13)

then $S_n \xrightarrow{D} Z$.

The following three corollaries are easy consequences of the three corollaries above and Lemma 1.2. They replace the majorizing measure condition (1.3) with the metric entropy condition (1.8).

Corollary 2.4. If (1.8), (2.10), and (2.11) hold, then $S_n \xrightarrow{D} Z$.

Corollary 2.5. If (1.8), (2.10), and (2.12) hold, then $S_n \xrightarrow{D} Z$.

Corollary 2.6. If (1.8), (2.10), and (2.13) hold, then $S_n \xrightarrow{D} Z$.

Notice that if a weak convergence result for random fields whose domain lacks a temporal component is of interest, a simpler version of Theorem 2.1 (and consequently Corollaries 2.1 through 2.6 that follow) is readily available. In particular, consider the random field

$$S_n(x) = \sum_{i \le \tau(n)} Y_{n,i}(x) \quad \text{for } x \in \mathbf{X},$$

where the $Y_{n,i}$'s are measurable with respect to an increasing filtration and $\tau(n)$ is a stopping time. Think of this as reducing the temporal component of the domain to a single value; i.e. $\mathbf{I} = \{1\}$. Conditions (2.8) and (2.9) of Theorem 2.1

would thus be simplified, and Corollaries 2.1 through 2.6 would hold as stated for $\mathbf{I} = \{1\}$ throughout.

3. Finite Dimensional Convergence and the Limiting Process. In the case in which the prelimiting random fields have independent "increments" in t, e.g. for $0<s<t, S_n(s, \cdot\ ), S_n(t, \cdot\ )-(S_n \cdot s,)$ are independent processes, the usual central limit theorem can be used to identify the limit law which will also, of course, have independent increments. However when a more complex 'temporal' dependence structure exists in S_n the identification of the distribution of the limiting becomes a serious problem. Here we state sufficient conditions for identification of a more general Gaussian limit by calling in general results for weak convergence of $\mathbf{R}^k$ -valued semimartingales. The following lemma is a reformulation of e.g. Thm 2.27 of Jacod and Shiryaev (1987).

Lemma 3.1. (Convergence of finite-dimensional distributions in **X**.) If for all $t \in \mathbf{I},\ \varepsilon \in (0,1], \underset{\sim}{a} = (a_1,\dots,a_k) \in \mathbf{R}^k$, and $\underset{\sim}{x} = (x_1,\dots x_k) \in \mathbf{X}^k$ for $k \geq 1$,

$$\sum_{1 \le i \le \tau(n,t)} P\left(\Big|\sum_{1 \le j \le k} a_j Y_{n,i}(x_j)\Big| > \varepsilon \,\Big|\, F_{n,i-1}\right) \xrightarrow{P} 0, \tag{3.1}$$

$$\sup_{t \in \mathbf{I}} \sum_{1 \le i \le \tau(n,t)} \left| E\left(\sum_{1 \le j \le k} a_j Y_{n,i}(x_j)\, 1_{\left[\left|\sum_{1 \le j \le k} a_j Y_{n,i}(x_j)\right| \le \varepsilon\right]} \,\Big|\, F_{n,i-1}\right)\right| \xrightarrow{P} 0, \tag{3.2}$$

and

$$\sum_{1 \le i \le \tau(n,t)} \left(E\left(\left(\sum_{1 \le j \le k} a_j Y_{n,i}(x_j)\right)^2 1_{\left[\left|\sum_{1 \le j \le k} a_j Y_{n,i}(x_j)\right| \le \varepsilon\right]} \,\Big|\, F_{n,i-1}\right) \right.$$

$$\left. - \left(E\left(\sum_{1 \le j \le k} a_j Y_{n,i}(x_j)\, 1_{\left[\left|\sum_{1 \le j \le k} a_j Y_{n,i}(x_j)\right| \le \varepsilon\right]} \,\Big|\, F_{n,i-1}\right)\right)^2 \right) \tag{3.3}$$

$$\xrightarrow{P} t \operatorname{Var}\left(\sum_{1 \le j \le k} a_j Z(x_j,1)\right),$$

where $\{Z(x,1) : x \in \mathbf{X}\}$ is a Gaussian process, then the finite-dimensional distributions in **X** of $\{S_n(\cdot,t) : t \in \mathbf{I}\}$ converge to those of $\{Z(\cdot, t) : t \in \mathbf{I}\}$ where $Z(\cdot,t)$ is a standard evolving Gaussian process.

The finite-dimensional convergence in **X** above gives us a candidate for our limiting process, as well as the canonical metric σ (see 2.1) on the product space

$\mathbf{X} \times \mathbf{I}$. Recall also the definitions of the metrics d, ρ, and τ given in (2.2), (2.3) and (2.4) respectively of section 2. The following lemma gives a proof of the continuity of the limiting distribution under the natural majorizing measure condition.

Lemma 3.2. If the mean-zero Gaussian random field $\{Z(x,1) : x \in \mathbf{X}\}$ is separable and d-continuous a.s., then the standard evolving Gaussian random field $\{Z(x,t) : x \in \mathbf{X}, t \in \mathbf{I}\}$ is σ-continuous and τ-continuous a.s..

Proof. First notice that for all $\delta > 0$, $x \in \mathbf{X}$ and $s \in \mathbf{I}$,

$$B_d(x, \delta/2) \times B_\rho(s, \delta/2) \subset B_\tau((x,s), \delta)\,. \tag{3.4}$$

From Theorem 1.1, $(\mathbf{X}, d)$ admits a majorizing measure ν satisfying

$$\lim_{\delta \to 0} \sup_{x \in \mathbf{X}} \int_0^\delta \ln^{1/2}(1/\nu(B_d(x,u)))\,du = 0.$$

Let m be the product measure $\nu \times \lambda$ on $\mathbf{X} \times \mathbf{I}$ where λ is Lebesgue measure. Then, setting $c = \sup_{x \in X}(\mathrm{Var}Z(x,1))^{1/2}$, it follows from (3.4) that

$$\int_0^\delta \ln^{1/2}(1/m\,(B_\tau((x,t),u)))\,du \le \int_0^\delta \ln^{1/2}(1/\lambda\ (B_\rho(t,u/2))\ \nu(B_d(x,u/2)))\,du$$

$$\le \int_0^\delta \ln^{1/2}(1/cu^{1/2})\,du + \int_0^\delta \ln^{1/2}(1/\nu(B_d(x,u/2)))\,du,$$

and thus ,

$$\lim_{\delta \to 0} \sup_{(x,t) \in \mathbf{X} \times \mathbf{I}} \int_0^\delta \ln^{1/2}(1/m(B_\tau((x,t),u)))\,du = 0. \tag{3.5}$$

Since $B_\tau((x,t),u) \subset B_\sigma((x,t),u)$, clearly (3.5) holds also with the metric τ replaced by σ. The result follows from Theorem 1.1. □

Again a parallel result is easily available for many general evolving Gaussian processes. (See comment following Theorem 2.1.)

4. Exponential Probability Bounds for Martingale Differences. This section contains some inequalities which play an essential part in this derivation of our tightness results. Throughout we assume that $\{Y_i : i \ge 1\}$ is a sequence of real-valued random variables on the probability space (Ω, F, P) which are measurable with respect to an increasing filtration $\{F_i : i \geq 1\}$ with $F_i \subset F$ for each i. We let $F_0 = \{\phi, \Omega\}$ denote the trivial σ-algebra.

The following proposition, a Bernstein-type inequality for martingales, due to Freedman (1973), is a basic ingredient of most of our probability bounds.

Proposition 4.1. . Suppose that $P(\sup_{i \geq 1} |Y_i| > a) = 0$ and $E(Y_i | F_{i-1}) = 0$. For $n \geq 1$, let $V_n = \sum_{1 \leq i \leq n} E(Y_i^2 | F_{i-1})$. Then, for all $\eta, \delta > 0$,

$$P\left[\sum_{1 \leq i \leq n} Y_i > \eta, V_n \leq \delta \text{ for some } n \geq 1\right] \leq [\delta / (a\eta + \delta)]^{(a\eta+\delta)/a^2} e^{\eta/a}$$

$\leq \exp\{-\eta^2/2\,(a\eta + \delta)\}$.

The following simple lemma proves useful. A variant of it may be found in AGOZ (1988).

Lemma 4.1. Let $\{c_i : i \geq 1\}$ be a decreasing sequence of non-negative real numbers. For any $b > 0$ and $n \geq 1$,

$$\sup_{a > 0} a \sum_{1 \leq i \leq n} c_i 1_{[a < c_i \leq b]} \leq 2 \sup_{1 \leq i \leq n} i c_i^2 1_{[c_i \leq b]}.$$

Proof: Notice first that for any $j \geq 1$,

$$\sum_{1 \leq i \leq j} c_i 1_{[c_i \leq b]} = \sum_{1 \leq i \leq j} (i^{1/2} c_i 1_{[c_i \leq b]}) i^{-1/2}$$

$$\leq \sup_{1 \leq i \leq j} i^{1/2} c_i 1_{[c_i \leq b]} \left(\sum_{1 \leq i \leq j} i^{-1/2}\right)$$

$$\leq 2j^{1/2} \sup_{1 \leq i \leq n} i^{1/2} c_i 1_{[c_i \leq b]}.$$

Fix $a > 0$, and let $j_a = \{j : c_j \geq a > c_{j+1}\}$. Then

$$a \sum_{1 \leq i \leq n} c_i 1_{[a < c_i \leq b]} = a \sum_{1 \leq i \leq j_a} c_i 1_{[a < c_i \leq b]}$$

$$\leq c_{j_a} 1_{[c_{j_a} \leq b]} \sum_{1 \leq i \leq j_a} c_i 1_{[c_i \leq b]}$$

$$\leq 2 j_a^{1/2} c_{j_a} 1_{[c_{j_a} \leq b]} \sup_{1 \leq i \leq n} i^{1/2} c_i 1_{[c_i \leq b]}$$

$$\le 2 \sup_{1\le j\le n} j^{1/2} c_j 1_{[c_j\le b]} \sup_{1\le i\le n} i^{1/2} c_i 1_{[c_i\le b]}$$

$$\le 2 \sup_{1\le j\le n} j c_j^2 1_{[c_j\le b]}. \qquad \square$$

The following lemma is an easy consequence of Freedman's Proposition (4.1).

Lemma 4.2. For any $\eta > \delta > 0$ and sequence of sets $\{A_i : i \ge 1\}$ with $A_i \in F_i$,

$$P\Big(\sum_{1\le i\le n} 1_{A_i}(Y_i) > \eta,\ \sum_{1\le i\le n} P(A_i | F_{i-1}) \le \delta \text{ for some n}\Big) \le (e\delta|\eta)^\eta.$$

Proof: Since $|1_{A_i}(Y_i) - P(A_i| F_{i-1})| \le 1$, $E(1_{A_i}(Y_i) - P(A_i| F_{i-1})) = 0$, and $E((1_{A_i}(Y_i) - P(A_i | F_{i-1}))^2 | F_{i-1}) \le P(A_i| F_{i-1})$, we may apply Proposition 4.1 above to see that

$$P\Big(\sum_{1\le i\le n} 1_{A_i}(Y_i) > \eta \text{ and } \sum_{1\le i\le n} P(A_i| F_{i-1}) \le \delta \text{ for some n}\Big)$$

$$\le P\Big(\sum_{1\le i\le n} (1_{A_i}(Y_i) - P(A_i| F_{i-1})) > \eta - \delta \text{ and } \sum_{1\le i\le n} P(A_i| F_{i-1}) \le \delta \text{ for some n}\Big)$$

$$\le (\delta/\eta)^\eta e^\eta. \qquad \square$$

The following lemma is essential in proving our tightness theorem. It is basically an extension of Lemma 2.16 of AGOZ (1988) to sums whose increments are martingale differences.

Lemma 4.3. Suppose that the Y_i 's are non-negative and let the set

$$\Lambda = \Big[\sup_{a>0} a^2 \sum_{i\ge 1} P(Y_i > a| F_{i-1}) \le \delta\Big].$$

For all $\eta > \delta$,

$$P\Big(\Big[\sup_{a>0} a^2 \sum_{1\le i\le n} Y_i 1_{[a<y_i\le b]} > 2\eta \text{ for some n}\Big] \cap \Lambda\Big)$$

$$\le (e\delta/\eta)^{\eta/b^2}/(1-e\delta/\eta).$$

Proof: Fix $n \ge 1$ and let $Y_{(i)}$ be the *decreasing* order statistics of $\{Y_1, ..., Y_n\}$. Using Lemma 4.1, we see that for $\eta > 0$

$$\Big[\sup_{a>0} a \sum_{1\le i\le n} Y_i 1_{[a<Y_i\le b]} > 2\eta\Big] \subset \Big[\sup_{1\le i\le n} iY_{(i)}^2 1_{[Y_{(i)}\le b]} > \eta\Big]$$

$$\subset \left[\sup_{1\le i\le n} iY^2_{(i)} 1_{[Y_{(i)}\le b]} > \eta\right]$$

$$= \bigcup_{\eta/b^2 < i \le n} \left[Y^2_{(i)} 1_{[Y_{(i)}\le b]} > \eta/i\right]$$

$$\subset \bigcup_{\eta/b^2 \le i < n} \left[\sum_{1\le j\le n} 1_{[Y_j > (\eta/i)^{1/2}]} \ge i\right].$$

Notice that, for each $i \geq 1$,

$$\Lambda \subset \left[(\eta/i) \sum_{1\le j\le n} P(Y_j > (\eta/i)^{1/2} | F_{j-1}) \le \delta \text{ for all n}\right]$$

$$= \left[\sum_{1\le j\le n} P(Y_j > (\eta/i)^{1/2} | F_{j-1}) \le \delta i\ \eta \text{ for all n}\right].$$

Thus, using Lemma 4.2, we see that

$$P\left(\left[\sup_{a>0} a \sum_{1\le i\le n} Y_i 1_{[a<Y_i\le b]} > 2\eta \text{ for some n}\right] \cap \Lambda\right)$$

$$\le \sum_{i>\eta/b^2} P\left(\left[\sum_{1\le j\le n} 1_{[Y_j > (\eta/i)^{1/2}]} \ge i \text{ for some n}\right] \cap \Lambda\right)$$

$$\le \sum_{i\le \eta/b^2} P\Bigg(\sum_{1\le j\le n} 1_{[Y_j > (\eta/i)^{1/2}]} \ge i \text{ and}$$

$$\sum_{1\le j\le n} P(Y_j > (\eta/i)^{1/2} | F_{j-1}) \le \delta i\ \eta \text{ for some n}\Bigg)$$

$$\le \sum_{i>\eta/b^2} (e\delta/\eta)^i$$

$$< (e\delta/\eta)^{\eta/b^2} / (1 - e\delta/\eta).$$

5. Tightness. We assume throughout this section that the metric space (**X**,d) admits a majorizing measure m for which

$$\lim_{\delta\to 0} \sup_{x\in X} \int_0^\delta ln^{1/2}(1/m(B_d(x,u)))\,du = 0. \tag{5.1}$$

Thus, in addition, we have, for each $\beta > 0$, a discrete majorizing measure μ on (**X**,d) for which (1.4) through (1.7) hold. Fix $\beta > 0$ and, in the notation of Lemma 1.1, set $\mathbf{X}_j = \{x_j : x \in \mathbf{X}\}$, so that $\mathbf{X}_j$ is a finite β^j net in **X**.

Let $\Delta_{n,i}^{(j)}(x) = \sup_{y : y_j = x_j} |Y_{n,i}(x_j) - Y_{n,i}(y)|$.

Thus $\Delta_{n,i}^{(j)}$ is a local modulus of continuity for $Y_{n,i}$ which is constant on the set $\{y : y_j = x\}$ for $x \in \mathbf{X}_j$. Due to the structure of the x_j's specified by (1.6), we may assume that $\Delta_{n,i}^{(j)}(x)$ decreases in j for each fixed x.

Before continuing, we introduce some technical definitions involving the discrete majorizing measure μ. Fix $0 < \beta < 1$. For $x \in \mathbf{X}$ and $j \geq 0$, set $m_j(x) = \beta^j \prod_{k \leq j} \mu(x_k)$. Define $m : X \times (0,1] \to (0,1]$ as follows:

$$m(x,u) = m_{j+1}(x) + (u - \beta^{j+1})(m_j(x) - m_{j+1}(x))/(\beta^j - \beta^{j+1})$$

for $\beta^{j+1} \leq u < \beta^j$. Notice that, for each fixed x, $m(x,\cdot)$ is a continuous and strictly increasing function with $m(x,u) = m(x_j, u)$ for $\beta^j \leq u \leq 1$. Also, since μ is a majorizing measure, we have $\sum_{x \in \mathbf{X}_j} m(x_j) \leq \beta^j$. Let $g : \mathbf{X} \times (0,1] \to \mathbf{R}^+$ be defined by

$$g(x,u) = ln(1/m(x,u)). \tag{5.2}$$

For each fixed x, the function $g(x,\cdot)$ is continuous and strictly decreasing with $g(x,u) = g(x_j, u)$ for $\beta^j \leq u \leq 1$. If the metric entropy integral condition (1.8) holds, we may assume that m(x,u) (and thus g(x,u)) depend only on u and not on x. Thus, under the assumption of (1.8) rather than (5.1) most of the conditions that follow can be stated much more simply.

The following sets $\Lambda_j^{(1)}$ and $\Lambda_j^{(2)}$ are respectively the random function analogues of the sets Λ of Lemma 4.3 and $[V_n \leq \delta]$ of Proposition 4.1. For each fixed n, denote the entire sequence $\{Y_{n,i} : i \geq 1\}$ by Y_n.

For $j \geq 0$, let

$$\Lambda_j^{(1)}(Y_n) = [\sup_{t \in \mathbf{I}} \sup_{a > 0} a^2 \sum_{1 \leq i \leq \tau(n,t)} P(\Delta_{n,i}^{(k)}(x) > a | F_{n,i-1}) > \beta^{2k} \tag{5.3}$$

$$\text{for some } x \in \mathbf{X}, k \geq j]$$

for $b > 0$

$$\Lambda_j^{(2)}(Y_n, b) = [\sup_{t \in \mathbf{I}} \sum_{1 \leq i \leq \tau(n,t)} E((Y_{n,i}(x_k) - Y_{n,i}(x_{k-1}))^2 \tag{5.4}$$

$$1_{[|Y_{n,i}(x_k) - Y_{n,i}(x_{k-1})| \le \beta^k / g^{1/2}(x,\beta^k)]} \Big| F_{n,i-1}) > b\beta^{2k}$$

for some $x \in \mathbf{X}$ and k > j],

and for c > 0

$$\Lambda_j^{(3)}(Y_{n,c}) = [\sup_{t \in \mathbf{I}} \sum_{1 \le i \le \tau(n,t)} P(|Y_{n,i}(x_k)| > \beta^j / 2g(x,\beta^j) | F_{n,i-1})$$

$$> cg(x,\beta^k) \text{ for some } x \in \mathbf{X}, k \ge j]. \tag{5.5}$$

Notice that if the increments, $Y_{n,i}$, are independent, then all three sets, $\Lambda^{(1)}, \Lambda^{(2)}$ and $\Lambda^{(3)}$ are non-random.

For a > 0, let

$$\phi(y,a) = -a1_{[y \le -a]} + y1_{[|y| < a]} + a1_{[y \ge a]}$$

and for $a : \mathbf{X} \to \mathbf{R}^+$, let

$$Y_{n,i}^{a(\cdot)}(x) = \phi(Y_{n,i}(x), a(x))$$

denote the truncated $Y_{n,i}$ process. Notice that we have allowed the truncation level to depend on x. Again, denote the entire process $\{Y_{n,i}^{a(\cdot)} : i \ge 1\}$ by $Y_n^{a(\cdot)}$, and let

$$S_n^{a(\cdot)}(x,t) = \sum_{1 \le i \le \tau(n,t)} (Y_{n,i}^{a(\cdot)}(x) - E(Y_{n,i}^{a(\cdot)}(x) | F_{n,i-1})) \tag{5.6}$$

denote the truncated and conditionally centered partial sum process.

We are ready to state our first tightness theorem.

Theorem 5.1 If (X,d) admits a majorizing measure satisfying (5.1),

$$\sup_{x \in \mathbf{X}, t \in \mathbf{I}} \sup_{1 \le i \le \tau(n,t)} |Y_{n,i}(x)| \xrightarrow{P} 0 \tag{5.7}$$

for all a > 0

$$\sup_{x \in \mathbf{X}, t \in \mathbf{I}} \sup_{1 \le i \le \tau(n,t)} P(|Y_{n,i}(x)| > a | F_{n,i-1}) \xrightarrow{P} 0 \tag{5.8}$$

and

$$\sup_{x \in \mathbf{X}, t \in \mathbf{I}} \Big| \sum_{1 \le i \le \tau(n,t)} E(Y_{n,i}(x) 1_{[|Y_{n,i}(x)| \le a]} \big| F_{n,i-1}) \Big| \xrightarrow{P} 0, \tag{5.9}$$

for some $b \le e^2 (e^2 - \beta^3)/\beta^4$,

$$\overline{\lim_j}\ \overline{\lim_n}\ P^* (\Lambda_j^{(1)} (Y_n) \cup \Lambda_j^{(2)} (Y_n, b)) = 0 \tag{5.10}$$

then, for all $\eta > 0$,

$$\overline{\lim_j}\ \overline{\lim_n}\ P^* (\sup_{x \in \mathbf{X}, t \in \mathbf{I}} |S_n(x,t) - S_n(x_j, t)| > \eta) = 0. \tag{5.11}$$

Theorem 5.1 is a corollary of the following theorem which has slightly weaker but more cumbersome hypotheses.

Theorem 5.2. If (X,d) admits a majorizing measure satisfying (5.1), for all $a > 0$ and $\varepsilon > 0$,

$$\overline{\lim_j}\ P^* (\sup_{x \in \mathbf{X}, t \in \mathbf{I}} \big|S_n(x,t) - S_n^a(x,t)\big| > \varepsilon) = 0 \tag{5.12}$$

and for some $b \le e^2(e^2 - \beta^3)/\beta^4$ and $c \le e^2(e^2 - \beta^3)/2\beta^4$,

$$\overline{\lim_j}\ \overline{\lim_n}\ P^* (\Lambda_j^{(1)} (Y_n) \cup \Lambda_j^{(2)} (Y_n, b) \cup \Lambda_j^{(3)} (Y_n, c)) = 0,$$

then for all $\eta > 0$, (5.11) holds.

Conditions (5.10) and (5.13) above should be viewed as the conditional analogue of the weak - L_2 and L_2 bracketing conditions for process with independent increments. (c.f. Ossiander (1986) and AGOZ (1988)).

The following lemma helps bridge the distance between Theorem 5.2 and 5.1.

Lemma 5.1. If (5.7) holds, and, for all $a > 0$, (5.8) and (5.9) hold, then for all $a > 0$ and $\varepsilon > 0$, (5.12) holds.

Proof: For any $x \in \mathbf{X}$, $t \in \mathbf{I}$, and $a > 0$,

$$\begin{aligned} &\big|S_n(x,t) - S_n^a(x,t)\big| \\ &\quad = \Big| \sum_{1 \le i \le \tau(n,t)} (Y_{n,i}(x) - Y_{n,i}^a(x) + E(Y_{n,i}^a(x) | F_{n,i-1})) \\ &\le \sum_{1 \le i \le \tau(n,t)} \big|Y_{n,i}(x) - Y_{n,i}^a(x)\big| \end{aligned}$$

$$+\left|\sum_{1\le i\le\tau(n,t)} E(Y^a_{n,i}(x)\,|\,F_{n,i-1})\right|$$

$$\le \sum_{1\le i\le\tau(n,t)} |Y_{n,i}(x)|\,1_{[|Y_{n,i}(x)|>a]}$$

$$+\left|\sum_{1\le i\le\tau(n,t)} E(Y_{n,i}(x)\,1_{[|Y_{n,i}(x)|\le a]}\,|F_{n,i-1})\right|$$

$$+a\sum_{1\le i\le\tau(n,t)} P(|Y_{n,i}(x)|>a|\,F_{n,i-1}).$$

Thus, for any a, $\eta > 0$,

$$P^*(\sup_{s\in \mathbf{X}, t\in \mathbf{I}} |S_n(x,t)-S^a_n(x,t)|>\eta)$$

$$\le P^*(\sup_{x\in\mathbf{X}}\sup_{1\le i\le\tau(n,t)} |Y_{n,i}(x)|>a)$$

$$+P^*(\sup_{x\in\mathbf{X}}\left|\sum_{1\le i\le\tau(n,t)} E(Y_{n,i}(x)\,1_{[|Y_{n,i}(x)|\le a]}\right|F_{n,i-1})\Big|>\eta/3)$$

$$+P^*(\sup_{t\in I}\sum_{1\le i\le\tau(n,t)} P(|Y_{n,i}(x)|>a|\,F_{n,i-1})>\eta/3a).$$ □

We now prove Theorem 5.1 using Lemma 5.1 and Theorem 5.2.

Proof of Theorem 5.1: Lemma 5.1 implies that under the hypothesis of Theorem 5.1, (5.12) holds. Note that $g(x,\beta^j)$ is constant on the set $\{y\in \mathbf{X}: y_j = x\}$ so that $\beta^j/2g(x,\beta^j)$ takes on a finite number of values which are bounded below away from 0. Also $g(x,\beta^k)$ increases in k, so

$$\Lambda^{(3)}_j(Y_n,c)\subset [\sup_{x\in \mathbf{X}, t\in \mathbf{I}}\sum_{1\le i\le\tau(n,t)} P(\,|Y_{n,i}(x)|\;>\min_{x\in\mathbf{X}_j}\beta^j/2g(x,\beta^j)\,|\,F_{n,i-1})$$

$$>c\min_{x\in\mathbf{X}_j} g(x,\beta^j)\,].$$

Condition (5.8) implies that the probability of this latter set goes to 0 for all $j\ge 0$ and $c>0$. The proof is completed by noting that

$$(P^*(\Lambda^{(1)}_j(Y_n)\cup\Lambda^{(2)}_j(Y_n,b)\cup\Lambda^{(3)}_j(Y_n,c))$$

$$\le P(\Lambda^{(1)}_j(Y_n)\cup\Lambda^{(2)}_j(Y_n,b))$$

$$+P^*\,(\Lambda_j^{(3)}\,(Y_n, c))\,. \qquad \square$$

Theorem 5.2 is, in turn, proved using the following result.

Theorem 5.3. If (**X** ,d) admits a majorizing measure satisfying (5.1) and for all $a > 0$ and $\varepsilon > 0$, (5.12) holds, then for all $\eta > 0$ and $b \le 2e^2(e^2 - \beta^3)/\beta^4$,

$$\overline{\lim_j}\ \overline{\lim_n}\ P^*([\sup_{x\in \mathbf{X}, t\in \mathbf{I}} |S_n(x,t) - S_n(x_j, t)| > \eta\,] \tag{5.14}$$

$$\cap\,(\Lambda_j^{(1)}\,(Y_n^{\beta^j/2g(\,\cdot\,,\beta^j)}\,)\cup\Lambda_j^{(2)}\left(Y_n^{\beta^j/2g(\,\cdot\,,\beta^j)}, b\,)\,)^c\,\right) = 0.$$

We use the following two lemmas in deriving Theorem 5.2 from Theorem 5.3.

Proof of Lemma 5.2: Fix $j \ge 0$ and to simplify notation, set $a_j(x) = \beta^j/2g^{1/2}(x,\beta^j)$. Clearly, since $a_j(x)$ only depends on x through x_j, $Y_{n,i}^{a_j(y)}(y)$ and $Y_{n,i}^{a_j(x_j)}(x_j)$ are truncated at the same level ($a_j(x_j)$) whenever $y_j = x_j$ and we have then

$$\left|Y_{n,i}^{a_j(y)}(y) - Y_{n,i}^{a_j(x_j)}(x_j)\right| \le |Y_{n,i}(y) - Y_{n,i}(x_j)|\,.$$

Thus, for $k \ge j$, using (1.6) of Lemma 1.1,

$$\sup_{y\,:\,y_k = x_k} \left|Y_{n,i}^{a_j(y)}(y) - Y_{n,i}^{a_j(x_j)}(x_j)\right| \le \Delta_{n,i}^{(k)}(x)\,. \qquad \square$$

Lemma 5.3. For all $c > 0$ and $j \ge 0$,

$$\Lambda_j^{(2)}\,(Y_n^{\beta^j/2g(\,\cdot\,,\beta^j)}, c) \subset \Lambda_j^{(2)}\,Y_n, c/2\,) \cup \Lambda_j^{(3)}\,(Y_n, c/4)\,.$$

Proof: Again set $a_j(x) = \beta^j/2g(\cdot,\beta^j)$ for ease of notation and note that $a_j(x)$ depends on x only through x_j. For any $a, b > 0$, we have

$$(Y_{n,i}^a(x_k) - Y_{n,i}^a(x_{k-1}))^2 1_{[|Y_{n,i}^a(x_k) - Y_{n,i}^a(x_{k-1})| \le b]}$$

$$\le (Y_{n,i}(x_k) - Y_{n,i}(x_{k-1}))^2 1_{[|Y_{n,i}(x_k) - Y_{n,i}(x_{k-1})| \le b]}$$

$$+\ (2b)^2\,(1_{[|Y_{n,i}(x_k)| > a]} + 1_{[|Y_{n,i}(x_{k-1})| > a]})$$

so, taking conditional expectations, summing, and taking suprema, we have, for any $j \ge 0$ and $c > 0$,

$$\Lambda_j^{(2)}(Y_n^{a_j(\,\cdot\,)}, c) \subset \Lambda_j^{(2)}(Y_n, c/2)$$

$$\cup\ [4a_k^2(x) \sum_{1 \le i \le \tau(n,t)} P(|Y_{n,i}(x_k)| > a_j(x) \mid F_{n,i-1})$$

$$> c\beta^{2k}/4 \text{ for some } x \in \mathbf{X} \text{ and } k \ge j]$$

$$\subset \Lambda_j^{(2)}(Y_n, c/2) \cup \Lambda_j^{(3)}(Y_n, c/4)\,. \qquad \square$$

Proof of Theorem 5.2: It follows from Lemmas 5.2 and 5.3 that for any $\eta > 0$, $j \ge 0$, and $b > 0$,

$$P^*(\sup_{x \in \mathbf{X}, t \in \mathbf{I}} |S_n(x,t) - S_n(x_j, t)| > \eta)$$

$$\le P^*\Big([\sup_{x \in \mathbf{X}, t \in \mathbf{I}} |S_n(x,t) - S_n(x_j,t)| > \eta] \cap (\Lambda_j^{(1)}(Y_n^{\beta^j/2g(\,\cdot\,,\beta^j)})$$

$$\cup \Lambda_j^{(2)}(Y_n^{\beta^j\,2g(\,\cdot\,,\beta^j)}, b))^c\Big)$$

$$+ P^*(\Lambda_n^{(1)}(Y_n) \cup \Lambda_n^{(2)}(Y_n, b/2) \cup \Lambda_n^{(3)}(Y_n, b/4))\,. \qquad \square$$

Theorem 5.3 depends in turn on the following proposition. The statement of the proposition employs the function $\gamma : \mathbf{X} \times (0,1] \to R^+$ defined by

$$\gamma(x, \delta) = \sum_{j \ge j_\delta + 1} \beta^j ln^{1/2}(1/m_j(x)), \tag{5.15}$$

where $j_\delta = \{j : \beta^{j+1} < \delta \le \beta^j\}$.

Proposition 5.1. If $(\mathbf{X}, d)$ admits a majorizing measure satisfying (5.1),

$$E(Y_{n,i}(x) \mid F_{n,i-1}) = 0 \text{ for } x \in \mathbf{X} \text{ and } i \ge 1$$

and, for some $\beta \in (0,1)$ and $j_0 \ge 0$,

$$P^*(\sup_{i \ge 1} |Y_{n,i}(x)| > \beta^{j_0}/2g^{1/2}(x, \beta^{j_0}) \text{ for any } x \in \mathbf{X}) = 0, \tag{5.16}$$

where g is as defined in (5.2), then for $b \le 2e^2(e^2 - \beta^3)/\beta^4$,

$$P^*([\sup_{t \in \mathbf{I}} |S_n(x,t) - S_n(x_{j_0}, t)| > (1 + 12e^2/\beta^5)\gamma(x, \beta^{j_0}) \text{ for some } x \in \mathbf{X}] \tag{5.17}$$

$$\cap \left(\Lambda_{j_0}^{(1)}(Y_n) \cup \Lambda_{j_0}^{(2)}(Y_n, b)\right)^c) \le c\beta^{j_0}$$

where $c = c_\beta$ is a universal constant depending only on β.

Proof of Theorem 5.3: For any $x \in \mathbf{X}, t \in \mathbf{I}$, $j \geq 0$, and $a: \mathbf{X} \to R^+$ which depends on x only through x_j,

$$|S_n(x,t) - S_n(x_j,t)| \leq \left|S_n(x,t) - S_n^{a(\,\cdot\,)}(x,t)\right| + \left|S_n(x_j,t) - S_n^{a(\,\cdot\,)}(x_j,t)\right| + \left|S_n^{a(\,\cdot\,)}(x,t) - S_n^{a(\,\cdot\,)}(x_j,t)\right|.$$

Thus, for any $j \geq 0$,

$$\sup_{x \in \mathbf{X}, t \in \mathbf{I}} |S_n(x,t) - S_n(x_j,t)| \leq \sup_{x \in \mathbf{X}, t \in \mathbf{I}} \left|S_n^{\beta^j/2g^{1/2}(x,\beta^j)}(x,t) - S_n^{\beta^j/2g^{1/2}(x,\beta^j)}(x_j,t)\right| + 2 \sup_{x \in \mathbf{X}, t \in \mathbf{I}} \left|S_n(x,t) - S_n^{\beta^j/2g^{1/2}(x,\beta^j)}(x,t)\right|.$$

Fix $\eta > 0$ and choose j_0 sufficiently large to have

$$(1 + 12e^2/\beta^5) \sup_{x \in \mathbf{X}} \gamma(x, \beta^{j_0}) < \eta/2.$$

(Recall from (1.5) of Lemma 1.1 that $\sup_{x \in \mathbf{X}} \gamma(x,\delta) \to 0$ as $\delta \to 0$.) Then for $j \geq j_0$

$$P^*([\sup_{x \in \mathbf{X}, t \in \mathbf{I}} (|S_n(x,t) - S_n(x_j,t)| > \eta\,] \cap (\Lambda_j^{(1)}(Y_n^{\beta^j/2g(\,\cdot\,,\beta^j)}) \cup \Lambda_j^{(2)}\left(Y_n^{\beta^j/2g(x,\beta^j)}, b\right)\Big)^c\Big)$$

$$\leq P^*([\sup_{t \in \mathbf{I}} |S_n(x,t) - S_n(x_j,t)| > (1 + 12e^2/\beta^5)\gamma(x,\beta^j) \text{ for some } x \in \mathbf{X}] \cap (\Lambda_j^{(1)}(Y_n^{\beta^j/2g(\,\cdot\,,\beta^j)}) \cup \Lambda_j^{(2)}(Y_n^{\beta^j/2g(\,\cdot\,,\beta^j)}, b))^c\,)$$

$$+ P^*\Big(\sup_{y \in \mathbf{X}_j, t \in \mathbf{I}, x \in \mathbf{X}} \ \sup_{x_j = y} \left|S_n(x,t) - S_n^{\beta^j/2g(y,\beta^j)}(x,t)\right| > \eta/4\Big)$$

$$\leq c\beta^j + o(1).$$ □

It remains to prove Proposition 5.1. It proceeds as follows. First we stratify $S_n(x,t)$ and $S_n(x_{j_0},t)$ by partitioning the sample space. Then, within each stratum, for a particular $j \geq j_0$ we compare $S_n(x,t)$ to $S_n(x_{j_0},t)$ in two stages. The first comparison gives rise to two remainder terms, one of which is contained in $\Lambda_{j_0}^{(1)}$ and the other of which is bounded in probability on $\Lambda_{j_0}^{(1)}$ using Lemma 4.3. The second comparison gives rise to a third remainder term, which we bound in probability on the set $\Lambda_{j_0}^{(2)}$.

Before we begin the proof, however, we need some technical lemmas which will be used in controlling the rate at which our approximations x_j converge to x. The following property of the function g defined in (5.2) above is of central importance.

Lemma 5.4. The function $g : \mathbf{X} \times (0,1] \to R^+$ satisfies

$$\int_0^\delta g^{1/2}(x,u)\,du < ((1-\beta)/\beta)\gamma(x,\delta) \tag{5.18}$$

as well as

$$\sup_{x\in\mathbf{X}} \int_0^\delta g^{1/2}(x,u)\,du < ((1-\beta)/\beta) \sup_{x\in\mathbf{X}} \gamma(x,\delta). \tag{5.19}$$

Proof: If $\beta^{j+1} < u \le \beta^j$, then, for any $x \in \mathbf{X}$, $g(x,u) < ln^{1/2}(1/m_{j+1}(x))$. Thus, again letting $j_\delta = \{j : \beta^{j+1} < \delta \le \beta^j\}$,

$$\begin{aligned}\int_0^\delta g^{1/2}(x,u)\,du &< \sum_{j\ge j_\delta} \int_{\beta^{j+1}}^{\beta^j} g^{1/2}(x,u)\,du \\ &< \sum_{j\ge j_\delta} (\beta^j - \beta^{j+1})\, lu^{1/2}(1/m_{j+1}(x)) \\ &= ((1-\beta)/\beta) \sum_{j\ge j_\delta+1} \beta^j ln^{1/2}(1/m_j(x)).\end{aligned}$$

□

Fix $\delta \in (0,1]$, and for each $x \in \mathbf{X}$ and $k \ge 0$, choose $\delta_k(x)$ to satisfy

$$\delta_k(x)/g^{1/2}(x,\delta_k(x)) = \delta\beta^k/g^{1/2}(x,\delta). \tag{5.20}$$

Lemma 5.5 For each x, the sequence $\{\delta_k : k \ge 0\}$ is a strictly decreasing sequence. In addition, if $\gamma(x,1) < \infty$, then $\lim_{k\to\infty} \delta_k(x) = 0$.

Proof: Since for each fixed x the sequence $\delta\beta^k/g^{1/2}(x,\delta)$ is strictly decreasing in k and the function $u/g^{1/2}(x,u)$ is strictly increasing in u, it is clear that $\delta_k(x)$ is a strictly decreasing sequence. From (5.18) we see that

$$g^{1/2}(x,u) \le (1-\beta)\gamma(x,1)/\beta u,$$

so

$$\delta_k(x) = \delta\beta^k g^{1/2}(x, \delta_k(x)) / g^{1/2}(x, \delta) \leq \delta(1-\beta)\beta^k \gamma(x, 1) / \beta\delta_k(x) g^{1/2}(x, \delta).$$

Thus $\delta_k(x) \leq (\delta(1-\beta)\gamma(x, 1) \; / \beta g^{1/2}(x, \delta))^{1/2} \beta^{k/2}$, so $\lim_{k \to \infty} \delta_k(x) = 0$ if $\gamma(x, 1) < \infty$. □

It is easy to see that $\delta_k(x)$ decreases more slowly than β^k.

Lemma 5.6. For each $x \in \mathbf{X}$ and $k \geq 0$,

$$\beta\delta_k(x) < \delta_{k+1}(x). \tag{5.21}$$

Proof: For any $x \in \mathbf{X}$

$$\beta\delta_k(x)/g^{1/2}(x,\delta_k(x)) = \delta_{k+1}(x)/g^{1/2}(x,\delta_{k+1}(x)) < \delta_{k+1}(x)/g^{1/2}(x,\delta_k(x)). \square$$

The following summability result is quite important.

Lemma 5.7. For $\delta_k(x)$ as defined in (5.20) above

$$\sum_{k \geq 0} \delta_k(x) g^{1/2}(x, \delta_k(x)) \leq (2/\beta)\gamma(x, \delta). \tag{5.22}$$

Proof: For any $x \in \mathbf{X}$ and $K > 0$,

$$(1-\beta) \sum_{0 \leq k \leq K} \delta_k(x) g^{1/2}(x, \delta_k(x))$$

$$= (1-\beta)\delta g^{1/2}(x, \delta) + \sum_{0 < k \leq K} \delta_k^2(x) \, (g^{1/2}(x, \delta_k(x))) / \delta_k(x)$$

$$- g^{1/2}(x, \delta_{k-1}(x)) / \delta_{k-1}(x))$$

$$= (1-\beta)\delta g^{1/2}(x, \delta) + \sum_{0 < k \leq K} \delta_k(x) g^{1/2}(x, \delta_k(x))$$

$$- \sum_{0 \leq k < K} \delta_{k+1}^2(x) g^{1/2}(x, \delta_k(x)) / \delta_k(x)$$

$$= -\beta\delta g^{1/2}(x, \delta) + \sum_{0 \leq k < K} (\delta_k(x) - \delta_{k+1}^2(x) / \delta_k(x) \;) g^{1/2}(x, \delta_k(x))$$

$$+ \delta_K(x) g^{1/2}(x, \delta_K(x))$$

$$\leq 2 \sum_{0 \leq k < K} (\delta_k(x) - \delta_{k+1}(x)) g^{1/2}(x, \delta_k(x)) + \delta_K(x) g^{1/2}(x, \delta_K(x))$$

$$\leq 2 \sum_{0 \leq k < K} \int_{\delta_{k+1}(x)}^{\delta_k(x)} g^{1/2}(x, u)\, du + \int_0^{\delta_K(x)} g^{1/2}(x, u)\, du$$

$$< 2\int_0^{\delta} g^{1/2}(x, u)\, du\ .$$

The first equality above follows from the recursion relation

$$\beta g^{1/2}(x, \delta_k(x))/\delta_k(x) = g^{1/2}(x, \delta_{k-1}(x))/\delta_{k-1}(x)\,.$$

The first inequality stems from dropping the leading negative term and observing that

$$\delta_k(x) - \delta_{k+1}^2(x)/\delta_k(x) = (\delta_k(x) - \delta_{k+1}(x))\,(\delta_k(x) + \delta_{k+1}(x))/\delta_k(x) \le$$

$$2(\delta_k(x) - \delta_{k-1}(x))\,.$$

To complete the proof use Lemma 5.4. □

Proof of Proposition 5.1: We partition the sample space using the following construction. Fix $j_0 \ge 0$, $\beta \in (0, 1)$, and set $\delta = \beta^{j_0}$. We use the sequence $\{\delta_k(x) : k \ge 0\}$ defined in (5.20) and discussed above. Note that $\delta_0(x) = \delta = \beta^{j_0}$ for all $x \in \mathbf{X}$. Also define the geometrically decreasing sequence

$$a_k(x) = \delta\beta^k/g^{1/2}(x, \delta) = \delta_k(x)/g^{1/2}(x, \delta_k(x))\,.$$

For each $x \in \mathbf{X}$, set $j_k(x) = \{j : \beta^{j+1} < \delta_k(x) \le \beta^j$. Again, note that $j_0(x) = j_0$ for all $x \in \mathbf{X}$. For $k \ge 0$ let

$$A_{n,i}^k(x) = [a_{k+1}(x) < \Delta_{n,i}^{(j_k(x))}(x) \le a_k(x)]\,,$$

$$\tilde{A}_{n,i}^k(x) = [a_{k+1}(x) < \Delta_{n,i}^{(j_k(x))}(x)]\,,$$

and

$$D_{n,i}^k(x) = (\bigcup_{0 \le j \le k} \tilde{A}_{n,i}^k(x))^c.$$

The following sets are defined iteratively as follows. Set

$$B_{n,i}^0(x) = \tilde{A}_{n,i}^0(x)$$

and for $k > 0$, set

$$B_{n,i}^k(x) = \tilde{A}_{n,i}^k(x) \backslash \bigcup_{0 \le j \le k-1} B_{n,i}^j(x)\,.$$

Notice that

$$B_{n,i}^k(x) = \tilde{A}_{n,i}^k(x) \backslash \bigcup_{0 \le j \le k-1} \tilde{A}_{n,i}^j(x)$$

$$= A^k_{n,i}(x) \setminus \bigcup_{0 \le j \le k-1} \tilde{A}^j_{n,i}(x)$$

$$\subset A^k_{n,i}(x),$$

$$B^k_{n,i}(x) = D^{k-1}_{n,i}(x) \setminus D^k_{n,i}(x), \tag{5.23}$$

and

$$D^k_{n,i}(x) \subset (\tilde{A}^k_{n,i}(x))^c. \tag{5.24}$$

Since the $Y_{n,i}(x)$ is bounded a.s. by $a_0(x)/2$ for all $n,i \geq 1$, each $\Delta^{(0)}_{n,i}(x)$ is bounded by $a_0(x)$ a.s.. Thus for any fixed x and $K \geq 0$, the collection of sets $\{D^k_{n,i}, B^k_{n,i} : 0 \leq k \leq K\}$ provide a disjoint partition of the sample space.

We now stratify S_n. For $k \geq 0$, set

$$S_{n,k}(x,t) = \sum_{1 \le i \le \tau(n,t)} \left(Y_{n,i}(x) 1_{B^k_{n,i}(x)} - E\left(Y_{n,i}(x) 1_{B^k_{n,i}(x)} \middle| F_{n,i-1} \right) \right)$$

and

$$T_{n,k}(x,t) = \sum_{1 \le i \le \tau(n,t)} \left(Y_{n,i}(x) 1_{D^k_{n,i}(x)} - E\left(Y_{n,i}(x) 1_{D^k_{n,i}(x)} \middle| F_{n,i-1} \right) \right).$$

Since $E(Y_{n,i}(x) | F_{n,i-1}) = 0$, for any $K \geq 0$,

$$S_n(x,t) = \sum_{0 \le k \le K} S_{n,k}(x,t) + T_{n,K}(x,t). \tag{5.25}$$

Each $S_{n,k}(x,t)$ is compared to its (approximately) $\delta_k(x)$-approximation $S_{n,k}(x_{j_k(x)}, t)$ as follows:

$$|S_{n,k}(x,t) - S_{n,k}(x_{j_k(x)}, t)| = \Big| \sum_{1 \le i \le \tau(n,t)} ((Y_{n,i}(x) - Y_{n,i}(x_{j_k(x)})) 1_{B^k_{n,i}(x)}$$

$$- E((Y_{n,i}(x) - Y_{n,i}(x_{j_k(x)})) 1_{B^k_{n,i}(x)} | F_{n,i-1})) \Big|$$

$$\le \sum_{1 \le i \le \tau(n,t)} E\left(\Delta^{(j_k(x))}_{n,i}(x) 1_{B^k_{n,i}(x)} \middle| F_{n,i-1} \right) + \sum_{1 \le i \le \tau(n,t)} \Delta^{(j_k(x))}_{n,i}(x) 1_{B^k_{n,i}(x)}$$

$$:= R^{(0)}_{n,k}(x,t) + R^{(1)}_{n,k}(x,t).$$

It is important to notice that, as both $\Delta^{(j_k(x))}_{n,i}(x)$ and , $B^k_{n,i}(x)$ depend on x only through $x_{j_k(x)}$, both $R^{(0)}_{n,k}$(x,t) and $R^{(1)}_{n,k}$(x,t) also depend on x only through $x_{j_k(x)}$. Likewise,

$$\left|T_{n,k}(x,t) - T_{n,k}(x_{j_k(x)},t)\right| \tag{5.27}$$

$$= \Big| \sum_{1\le i\le \tau(n,t)} ((Y_{n,i}(x) - Y_{n,i}(x_{j_k(x)}))1_{D^k_{n,i}(x)}$$

$$-E((Y_{n,i}(x) - Y_{n,i}(x_{j_k(x)}))1_{D^k_{n,i}(x)}\big| F_{n,i-1}))$$

$$\le \sum_{1\le i\le \tau(n,t)} \left(\Delta^{j_k(x)}_{n,i}(x)\,1_{D^k_{n,i}(x)} + E\left(\Delta^{j_k(x)}_{n,i}(x)\,1_{D^k_{n,i}(x)}\big| F_{n,i-1}\right)\right)$$

$$\le 2\tau(n,t)\,a_{k+1}(x).$$

In turn we stratify $S_n(x_{j_0},t)$ and compare each stratum to $S_{n,k}(x_{j_k(x)},t)$ for a particular k. Let

$$S^{(0)}_{n,k}(x,t) = \sum_{1\le i\le \tau(n,t)} \left(Y_{n,i}(x_{j_0})\,1_{B^k_{n,i}(x)} - E\left(Y_{n,i}(x_{j_0})\,1_{B^k_{n,i}(x)}\big| F_{n,i-1}\right)\right),$$

$$T^{(0)}_{n,k}(x,t) = \sum_{1\le i\le \tau(n,t)} \left(Y_{n,i}(x_{j_0})1_{D^k_{n,i}(x)} - E\left(Y_{n,i}(x_{j_0})\,1_{D^k_{n,i}(x)}\big| F_{n,i-1}\right)\right),$$

and note that for any $K \ge 0$, and $x \in \mathbf{X}$

$$S_n(x_{j_0},t) = \sum_{0\le k\le K} S^{(0)}_{n,k}(x,t) + T^{(0)}_{n,K}(x,t).$$

Then, for any $K \ge 0$,

$$S_n(x_{j_0},t) - \left(\sum_{0\le k\le K} S_{n,k}(x_{j_k(x)},t) + T_{n,K}(x_{j_K(x)},t)\right) \tag{5.28}$$

$$= \sum_{0\le k\le K}\left(S^{(0)}_{n,k}(x,t) - S_{n,k}(x_{j_k(x)},t)\right) + \left(T^{(0)}_{n,K}(x,t) - T_{n,K}(x_{j_K(x)},t)\right)$$

$$= \sum_{0<k\le K}\ \sum_{1\le i\le \tau(n,t)} ((Y_{n,i}(x_{j_0(x)}) - Y_{n,i}(x_{j_k(x)}))1_{B^k_{n,i}(x)}$$

$$-E((Y_{n,i}(x_{j_0(x)}) - Y_{n,i}(x_{j_k(x)}))1_{B^k_{n,i}(x)}\big| F_{n,i-1}))$$

$$+ \sum_{1\le i\le \tau(n,t)} ((Y_{n,i}(x_{j_0(x)}) - Y_{n,i}(x_{j_K(x)}))1_{D^K_{n,i}(x)}$$

$$-E((Y_{n,i}(x_{j_0(x)}) - Y_{n,i}(x_{j_K(x)}))1_{D^K_{n,i}(x)}\big| F_{n,i-1}))$$

$$= \sum_{0<k\le K} \sum_{1\le i\le \tau(n,t)} ((Y_{n,i}(x_{j_{k-1}(x)}) - Y_{n,i}(x_{j_k(x)}))1_{D_{n,i}^{k-1}(x)\setminus D_{n,i}^{K}(x)}$$

$$- E((Y_{n,i}(x_{j_{k-1}(x)}) - Y_{n,i}(x_{j_k(x)}))1_{D_{n,i}^{k-1}(x)\setminus D_{n,i}^{K}(x)} \mid F_{n,i-1}))$$

$$+ \sum_{0<k\le K} \sum_{1\le i\le \tau(n,t)} ((Y_{n,i}(x_{j_{k-1}(x)}) - Y_{n,i}(x_{j_k(x)}))1_{D_{n,i}^{K}(x)}$$

$$- E((Y_{n,i}(x_{j_{k-1}(x)}) - Y_{n,i}(x_{j_k(x)}))1_{D_{n,i}^{K}(x)} \mid F_{n,i-1}))$$

$$= \sum_{0<k\le K} \sum_{1\le i\le \tau(n,t)} ((Y_{n,i}(x_{j_{k-1}(x)}) - Y_{n,i}(x_{j_k(x)}))1_{D_{n,i}^{k-1}(x)}$$

$$- E((Y_{n,i}(x_{j_{k-1}(x)}) - Y_{n,i}(x_{j_k(x)}))1_{D_{n,i}^{k-1}(x)} \mid F_{n,i-1})).$$

$$:= \sum_{0<k\le K} R_{n,k}^{(2)}(x,t)$$

In deriving the third equality, we interchange summation and note that

$$\sum_{0<k\le K} (y_0 - y_k)1_{B_k} = \sum_{0<k\le K} \sum_{0<j\le k} (y_{j-1} - y_j)1_{B_k}$$

$$= \sum_{0<j\le K} (y_{j-1} - y_j) \sum_{j\le k\le K} 1_{B_k}$$

$$= \sum_{0<j\le K} (y_{j-1} - y_j) 1_{\bigcup_{j\le k\le K} B_k}.$$

We then apply (5.23) to see that $\bigcup_{j\le k\le K} B_{n,i}^{k}(x) = D_{n,i}^{j-1}(x)\setminus D_{n,i}^{K}(x)$. Now for $t\in \mathbf{I}$, set

$$K(n,t) = \inf\{k : k \ge \sup_{x\in \mathbf{X}} ln(\gamma(x,\delta)\, g^{1/2}(x,\delta)\,/2\delta\beta\tau(n,t)\,)/ln\beta\}.$$

K(n,t) is finite a.s., since condition (5.1) (through Lemmas 1.1 and 5.4) insures that both $\gamma(\cdot,\delta)$ and g $(\cdot,\delta)$ are bounded on **X**. Now combine the comparisons (5.25), (5.26), (5.27) and (5.28) derived above to arrive at the following:

$$|S_n(x,t) - S_n(x_{j_0},t)| \tag{5.29}$$

$$\le |\sum_{0\le k\le K(n,t)} (S_{n,k}(x,t) - S_{n,k}(x_{j_k(x)},t))|$$

$$+ |T_{n,K(n,t)}(x,t) - T_{n,K(n,t)}(x_{j_{K(n,t)}}(x),t)|$$

$$+\left|\sum_{0\le k\le K(n,t)} S_{n,k}(x_{j_k(x)},t)+T_{n,K(n,t)}(x_{j_{K(n,t)}},t)-S_n(x_{j_0},t)\right|$$

$$\le \sum_{0<k\le K(n,t)} (R^{(0)}_{n,k}(x,t)+R^{(1)}_{n,k}(x,t))$$

$$+2\tau(n,t)\,a_{K(n,t)+1}(x)+\sum_{0<k\le K(n,t)} \left|R^{(2)}_{n,k}(x,t)\right|$$

$$\le \sum_{0<k\le K(n,t)} (R^{(0)}_{n,k}(x,t)+R^{(1)}_{n,k}(x,t))+\sum_{0<k\le K(n,t)} \left|R^{2}_{n,k}(x,t)\right|+\gamma(x,\delta).$$

For $k \geq 0$, set

$$\eta_k(x) = (2e^2/\beta^4)\,\delta_k(x)\,g^{1/2}(x,\delta_k(x)).$$

Then, from Lemma 5.7,

$$\sum_{k\ge 0}\eta_k(x) < (4e^2/\beta^5)\,\gamma(x,\delta).$$

Thus, for any $b > 0$, from (5.29),

$$P^*([\sup_{t\in\mathbf{I}}|S_n(x,t)-S_n(x_{j_0},t)| > (1+12e^2/\beta^5)\,\gamma(x,\delta) \text{ for some } x\in\mathbf{X}] \tag{5.30}$$
$$\cap\left(\Lambda^{(1)}_{j_0(Y_n)}\cup\Lambda^{(2)}_{j_0}(Y_n,b)\right)^c)$$

$$\le P^*([R^{(0)}_{n,k}(x) > \eta_k(x) \text{ for some } x\in\mathbf{X} \text{ and } 0\le k\le K(n,t)] \cap \left(\Lambda^{(1)}_{j_0}(Y_n)\right)^c)$$

$$+P^*([R^{(1)}_{n,k}(x) > \eta_k(x) \text{ for some } x\in\mathbf{X} \text{ and } 0\le k\le K(n,t)] \cap \left(\Lambda^{(1)}_{j_0}(Y_n)\right)^c)$$

$$+P^*([R^{(2)}_{n,k}(x) > \eta_k(x) \text{ for some } x\in\mathbf{X} \text{ and } 0\le k\le K(n,t)] \cap \left(\Lambda^{(2)}_{j_0}(Y_n,b)\right)^c).$$

The set where the first remainder is large is actually contained in $\Lambda^{(1)}_{j_0}$. For any $x \in \mathbf{X}$ and $t \in \mathbf{I}$, we have

$$R^{(0)}_{n,k}(x,t) \le a_k(x) \sum_{1\le i\le \tau(n,t)} P(\Delta^{(j_k(x))}_{n,i}(x) > a_{k+1}(x)\,|\,F_{n,i-1}),$$

so that

$$\bigcup_{0\le k\le K(n,t)} [R^{(0)}_{n,k}(x,t) > \eta_k(x)]$$

$$\subset \bigcup_{k\ge 0}\left[a^2_{k+1}(x)\sum_{1\le i\le\tau(n,t)} P(\Delta^{j_k(x)}_{n,i} > a_{k+1}(x)\,|\,F_{n,i-1}) > 2e^2\delta^2_k(x)\ /\beta^2\ \right]$$

$$\subset \bigcup_{k\geq 0} [\sup_{a>0} a^2 \sum_{1\leq i\leq \tau(n,t)} P(\Delta_{n,i}^{j_k(x)}(x) > a | F_{n,i-1}) > 2e^2\beta^{2j_k(x)}]$$

$$\subset \bigcup_{j\geq j_0} [\sup_{a>0} a^2 \sum_{1\leq i\leq \tau(n,t)} P(\Delta_{n,i}^{j}(x_j) > a | F_{n,i-1}) > \beta^{2j}]$$

$$\subset \Delta_{j_0}^{(1)}(Y_n).$$

Bounding the second remainder term in probability is more complicated. For $j \geq j_0$, set $b_j(x) = \beta^j/g^{1/2}(x,\beta^j)$. From the definition of $j_k(x)$ and $a_k(x)$ observe that $\delta_k(x) \leq \beta^{j_k(x)}$ from Lemma 5.6 and $a_k(x) \leq b_{j_k(x)}(x)$, so for any fixed $x \in \mathbf{X}$ and $t \in \mathbf{I}$,

$$\bigcup_{k\geq 0} [R_{n,k}^{(1)}(x,t) > \eta_k(x)]$$

$$\subset \bigcup_{k\geq 0} [a_{k+1}(x) \sum_{1\leq i\leq \tau(n,t)} \Delta_{n,i}^{(j_k(x))}(x) 1_{[a_{k+1}(x) < \Delta_{n,i}^{(j_k)(x)}(x) \leq a_k(x)]} > 2e^2\delta_k^2(x)/\beta^3]$$

$$\subset \bigcup_{k\geq 0} [\sup_{a>0} a \sum_{1\leq i\leq \tau(n,t)} \Delta_{n,i}^{(j_k(x))}(x) 1_{\left[a < \Delta_{n,i}^{(j_k(x))}(x) \leq b_{j_k(x)}(x)\right]} > 2e^2\beta^{2j_k(x)-1}]$$

$$\subset \bigcup_{j\geq j_0} [\sup_{a>0} a \sum_{1\leq i\leq \tau(n,t)} \Delta_{n,i}^{(j)}(x_j) 1_{[a < \Delta_{n,i}^{(j)}(x_j) \leq b_j(x_j)]} > 2e^2\beta^{2j-1}].$$

Thus, using the above, Lemma 4.3, and the definitions of the functions g(·,·) and m(·,·), we see that

$$P^*([R_{n,k}^{(1)}(x,t) > \eta_k(x) \text{ for some } x \in \mathbf{X}, t \in \mathbf{I}, \text{ and } k \geq 0] \cap \left(\Lambda_{j_0}^{(1)}(Y_n)\right)^c) \quad (5.32)$$

$$\leq \sum_{j\geq j_0} \sum_{x\in \mathbf{X}_j} P^*([\sup_{a>0} a \sum_{1\leq i\leq \tau(n,t)} \Delta_{n,i}^{(j)}(x) 1_{[a<\Delta_{n,i}^{(j)}(x) \leq b_j(x)]}$$

$$> 2e^2\beta^{2j-1} \text{ for some } t \in \mathbf{I}] \cup \left(\Lambda_{j_0}^{(1)}(Y_n)\right)^c)$$

$$\leq \sum_{j\geq j_0} \sum_{x\in \mathbf{X}_j} (\beta/e)^{e^2 g(x,\beta^j)/\beta}/(1-\beta/e)$$

$$\leq \sum_{j\geq j_0} \sum_{x\in \mathbf{X}_j} m(x,\beta^j)/(1-\beta/e)$$

$$\leq \beta^{j_0}/(1-\beta/e).$$

Finally, $R^{(2)}_{n,k}(x,t)$ is made up of the conditionally centered increments

$$(Y_{n,i}(x_{j_{k-1}(x)}) - Y_{n,i}(x_{j_k(x)}))\; 1_{D^{k-1}_{n,i}(x)}$$
$$-E((Y_{n,i}(x_{j_{k-1}(x)}) - Y_{n,i}(x_{j_k(x)}))1_{D^{k-1}_{n,i}(x)}\big| F_{n,i-1})$$

which depend on x only through $x_{j_k(x)}$. From (5.24), $D^{k-1}_{n,i}(x) \subset [\Delta^{(k-1)}_{n,i}(x) \le a_k(x)]$, and from Lemma 5.6, we know $\delta_k(x) > \beta\delta_{k-1}(x)$, so either $j_{k-1}(x) = j_k(x) + 1$ or $j_{k-1}(x) = j_k(x)$. Thus

$$Y_{n,i}(x_{j_{k-1}(x)}) - Y_{n,i}(x_{j_k(x)}) = \begin{cases} Y_{n,i}(x_{j_k(x)-1}) - Y_{n,i}(x_{j_k(x)}) & if\ j_{k-1}(x) \ne j_k(x) \\ 0 & ow \end{cases}. \tag{5.33}$$

Furthermore, if $j_{k-1}(x) \ne j_k(x)$, on the set $D^{k-1}_{n,i}(x)$ we have

$$\left|Y_{n,i}(x_{j_{k-1}(x)}) - Y_{n,i}(x_{j_k(x)})\right| \le a_k(x) = \delta_k(x)/g^{1/2}(x, \delta_k(x)) \le \beta^{j_k(x)}/g^{1/2}(x, \beta^{j_k(x)}).$$

Set $k_j(x) = \{k : \delta_k(x) \le \beta^j < \delta_{k-1}(x)\}$, so $k_j(x) = k_j(x_j)$, and notice that again from Lemma 5.6 we have $\delta_{k_j(x)}(x) \ge \beta\delta_{k_j(x)-1}(x) > \beta^{j+1}$, so that

$$\eta_{k_j(x)} = (2e^2/\beta^4)\,\delta_{k_j(x)}(x)\,g^{1/2}(x, \delta_{k_j(x)}(x)) \;\ge 2e^2\beta^{j-3}g^{1/2}(x_j, \beta^j), \tag{5.34}$$

as well as $a_{k_j(x)} \le \beta^j/g^{1/2}(x, \beta^j)$. Then, for any fixed $x \in \mathbf{X}, t \in \mathbf{I}$, from (5.33) and (5.34),

$$\bigcup_{k>0} [|R^{(2)}_{n,k}(x,t)| > \eta_k(x)] = \bigcup_{j>j_0} \left[|R^{(2)}_{n,k_j(x)}(x,t)| > \eta_{k_j(x)}(x)\right]$$
$$\subset \bigcup_{j>j_0} \left[|R^{(2)}_{n,k_j(x)}(x_j,t)| > 2e^2\beta^{j-3}g^{1/2}(x,\beta^j)\right]$$

and

$$[|R^{(2)}_{n,k}(x,t)| > \eta_k(x) \text{ for some } x \in \mathbf{X}, t \in \mathbf{I}, \text{and } k > 0]$$
$$\subset \bigcup_{j>j_0} \bigcup_{x \in \mathbf{X}_j} [|R^{(2)}_{n-k_j(x)}(x,t)| > 2e^2\beta^{j-3}g^{1/2}(x,\beta^j) \text{ for some } t \in \mathbf{I}]. \tag{5.35}$$

Set

$$E_{n,j}(x,t) = [$$

$$\sum_{1 \le i \le \tau(n,t)} E((Y_{n,i}(x_{j-1}) - Y_{n,i}(x_j))^2 1_{[|Y_{n,i}(x_{j-1}) - Y_{n,i}(x_j)| \le a_j(x)]}$$

$$| F_{n,i-1}) \le 2(e^2 - \beta^3) e^2 \beta^{2j-6}]$$

and note that $E_{n,j}(x,t) \supset (\Lambda^{(2)}(Y_n, b))^c$ for $b \le 2e^2(e^2-\beta^3)/\beta^4$. Then, from (5.35), Proposition 4.1, and the definition of g(·,·), we have

$$P^*([|R^{(2)}_{n,k}(x,t)| > \eta_k(x) \text{ for some } x \in \mathbf{X}, t \in \mathbf{I}, \text{ and } k > 0] \cap (\Lambda^2(Y_n, b))^c) \tag{5.36}$$

$$\le \sum_{j > j_0} \sum_{x \in \mathbf{X}_j} P^* \Big([|R^{(2)}_{n,k_j(x)}(x,t)| > 2e^2 \beta^{j-3} g^{1/2}(x, \beta^j)] \cap E_{n,j}(x,t)$$

$$\text{for some } t \in \mathbf{I} \Big)$$

$$\le \sum_{j > j_0} \sum_{x \in \mathbf{X}_j} 2 \exp\{-4e^4 \beta^{2j-6} g(x, \beta^j) / 2(2e^2\beta^{2j-3} + 2(e^2 - \beta^3) e^2 \beta^{2j-6}) \}$$

$$\le \sum_{j > j_0} \sum_{x \in \mathbf{X}_j} 2 \exp\{-g(x, \beta^j)\}$$

$$\le 2\beta^{j_0+1}/(1-\beta).$$

Combining (5.30), (5.31), (5.32) and (5.36), we have, for $b \le 2e^2(e^2-\beta^3)/\beta^4$,

$$P^*([\sup_{t \in \mathbf{I}} |S_n(x,t) - S_n(x_{j_0}, t)| > (1 + 12e^2/\beta^5)\gamma(x, \beta^{j_0}) \text{ for some } x \in \mathbf{X}]$$

$$\cap \Big(\Lambda^{(1)}_{j_0}(Y_n) \cup \Lambda^{(2)}_{j_0}(Y_n, b) \Big)^c \Big)$$

$$\le \beta^{j_0}(1 + \beta - 2\beta^2/e)/(1-\beta)(1-\beta/e),$$

which completes the proof of (5.17). □

References

[1] Andersen, N.T. (1986). Lectures on Talagrand's result about continuous Gaussian processes. Texas A&M preprint.

[2] Andersen, N.T. and Dobrić, V. (1987). The central limit theorem for stochastic processes. *Ann. Prob. 15*, 164-177.

[3] Andersen, N.T., Giné, E., Ossiander, M., and Zinn, J. (1988), The central limit theorem and a law of the iterated logarithm for empirical processes under local conditions, *Prob. Theory Rel. 77*, 271-305.

[4] Dudley, R.M. (1978), Central limit theorems for empirical measures, *Ann. Prob. 6*, 899-929.

[5] Ethier, S.N. and Kurtz, T.G. (1986), Markov processes: characterization and convergence. New York: Wiley.

[6] Fernique, X. (1974), Regularité des trajectoires des functions aléatoires gaussiennes, *Lect. Notes in Math. 480*, 1-96, Springer, Berlin.

[7] Freedman, D.A. (1973), On tail probabilities for martingales, *Ann. Prob. 3*, 100-118.

[8] Gaenssler, P. and Schneemeier, W. (1986), On functional limit theorems for a class of stochastic processes indexed by pseudo-metric parameter spaces (with applications to empirical processes), Mathematisches Institut, Lehrstuhl fuer Math Stochastik, Preprint No. 35, 1986.

[9] Goldie, C.M. and Greenwood, P.E. (1986a), Characterizations of set-indexed Brownian motion and associated conditions for finite-dimensional convergence, *Ann. Prob. 14*, 802-816.

[10] Goldie C.M. and Greenwood, P.E. (1986b), Variance of set-indexed sums of mixing random variables and weak convergence of set-indexed processes, *Ann Prob. 14*, 817-839.

[11] acod, J. and Shiryaev, A.N. (1987), Limit theorems for stochastic processes, Springer-Verlag.

[12] Koul, H.L. (1989), A weak convergence result useful in robust autoregression. Michigan State Univ. Tech. Report, 21 pgs.

[13] Ledoux, M. and Talagrand, M. (1989), Comparison theorems, random geometry and some limit theorems for empirical processes, *Ann Prob. 17*, 596-631.

[14] Liggett, T.M. (1985), Interacting particle systems. New York: Springer Verlag.

[15] Ossiander, M. (1985), Weak convergence and a law of the iterated logarithm for processes indexed by points in a metric space, Ph.D. thesis, University of Washington.

[16] Ossiander, M. (1986), A central limit theorem under metric entropy with L_2 bracketing, *Ann Prob. 15*, 897-919.

[17] Preston, C. (1972), Continuity properties of some Gaussian processes, *Ann. Math. Statist. 43*, 285-292.

[18] Talagrand, M. (1987), Regularity of Gaussian processes, *Acta. Math. 159*, 99-149.

[19] Walsh, J.B. (1984), An introduction to stochastic partial differential equations, École d' Éte de Probability's de Saint-Fleur XIV, Lecture Notes in Math., Springer *1180*, édité par P.L. Hennequin.

THE TWO-LOCUS ANCESTRAL GRAPH

R. C. Griffiths
Mathematics Department
Monash University

Abstract

In a population genetics two-locus model with recombination an offspring has either a single parent gene, or is a recombinant from two parent genes. The number of ancestors, backward in time, of a sample of genes can thus decrease or increase and is found to be a birth and death process. Instead of a one-locus ancestral tree the ancestral paths of a sample of gene pairs are described by a graph with leaves as the sample genes and an eventual common ancestor where all paths from the leaves lead. In this paper properties of the two-locus ancestral graph and the two marginal ancestral trees are studied.

1. Introduction. In population genetics the ancestral tree of a sample of genes plays an important role in a probabilistic description of the sample. Kingman (1982) studies the *coalescent process* which describes the relationship between a sample of genes and their ancestors. The ancestral tree of a sample in a large population can be described by a death process where ancestors coalesce at a rate of $r(r-1)/2$. A tree is constructed by beginning with leaves at the n edges and joining edges where ancestors coalesce, keeping edge lengths proportional to the times between coalescence. The root is the first common ancestor of the sample. Mutations occur (to ancestors) along the edges of the tree according to a Poisson process of rate $\theta/2$ and determine the allelic configuration in the sample. The infinitely-many-alleles model is characterized by every mutation producing a type entirely new to the population. A nice review article on ancestral trees is Tavare (1984).

In a multi-locus model with recombination the relationship between a sample of genes and their ancestors is complicated because of recombination splitting up genes.

A gene in the two-locus model is described by an ordered pair $(x, y) \in [0,1] \times [0,1]$. The allele types at locus A and locus B are x and y. Stochastic evolution of the population is described by a discrete-time Wright-Fisher model. There are a fixed number of M genes in each generation. Genes in generation $\tau+1$ are produced from those in generation τ in the following way. Choose a single parent at random from the preceeding generation with probability $1 - r_M$. With probability r_M, the parent is formed by recombining two genes chosen at random from the preceeding generation; if the first chosen is (α, β) and the second chosen is (γ, δ), then the parent is of type (α, δ). If the parents type is (x, y) then the offspring's type is (x', y'), where $x' = x$ with probability $1 - u_A$ or $x' = Z$, a uniform random variable on $[0, 1]$, with probability u_A, and similarly for y'. (Z is considered a member of an independent, identically distributed sequence of random variables where different mutations are represented by distinct random variables.) The labeling by uniform random variables is just a convenient way of achieving unique labels for mutations in the infinitely-many-alleles model.

A diffusion time scaling is to measure time in units of M generations and let $M \to \infty$ while $\theta_A = 2Mu_A$, $\theta_B = 2Mu_B$, $\rho = 2Mr_M$ are held constant. In Ethier and Griffiths (1990a) the Wright-Fisher model is set up as a measure-valued Markov process which converges to a measure-valued diffusion. An atom of the measure at (x, y) represents the relative frequency of that type in the population. The convergence is robust under a number of different models (eg. Moran model). Our interest here is in describing the process of a sample of n genes' ancestors backwards in time, rather than in convergence of the process. Events occurring between the sample and the sample's parents in the preceeding generation (to order M^{-1} in probability) are: mutation to a gene in the sample at the A locus, with probability nu_A; mutation to a gene in the sample at the B locus, with probability nu_B; coalescence, when the sample has $n - 1$ parents, with probability $n(n - 1)/2M$; and recombination, when one gene in the sample is constructed from two in the preceeding generation, with probability nr_M.

Let $n(t)$ be the number of ancestors at time t backwards of a sample of $n(0) = n$ genes in the limit process. This includes recombinant ancestors. Then $\{n(t); t \geq 0\}$ is a birth and death process with rates $\lambda_k = k\rho/2$, $\mu_k = k(k - 1)/2$. Let W_n be the waiting time to first visit state 1. W_n has a proper distribution because of the quadratic death and linear birth rates. It is convenient to stop the process at this time, since the genetic composition of the sample is determined by then. If the process was continued it would have a modified Poisson stationary distribution $\{(e^{\rho} - 1)^{-1} \rho^j/j!, j = 1,2,...\}$.

2. Two-locus Ancestral Graph. A *two-locus ancestral graph* is drawn proportional to a real-time scale starting from n end edges, then joining edges at a vertex when two ancestors of the sample coalesce, or appending two edges at a

Figure 1
Two-locus ancestal graph

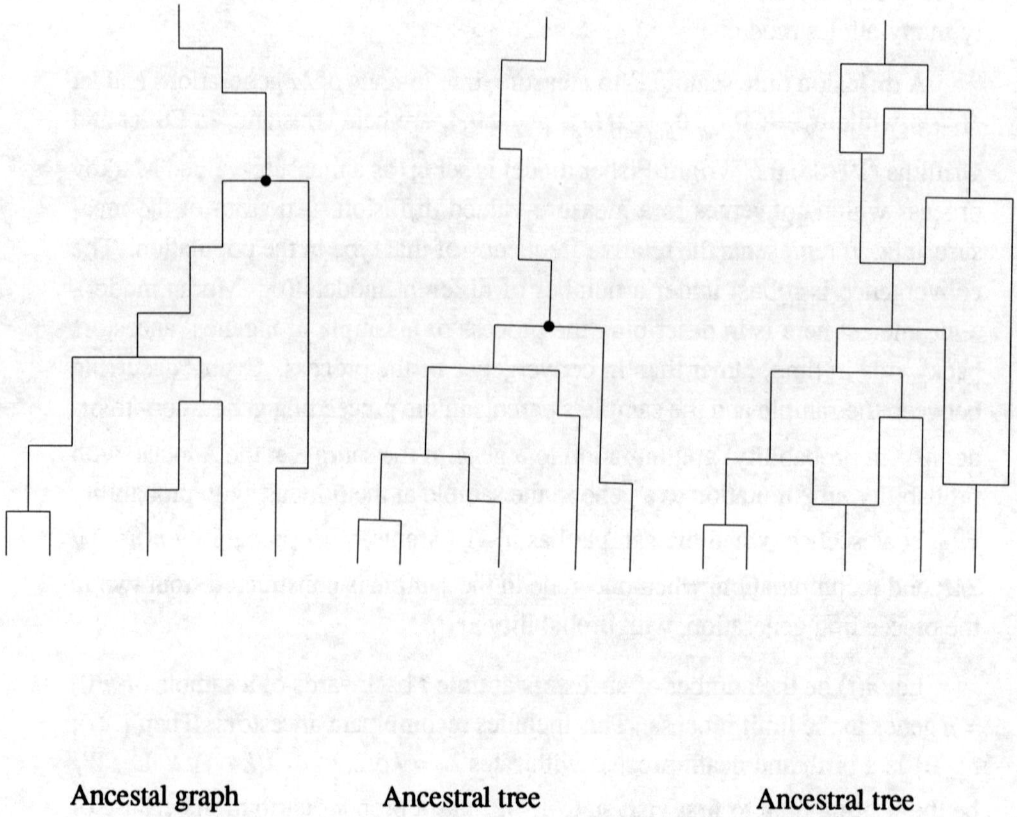

Ancestal graph

Ancestral tree Locus A

Ancestral tree Locus B

vertex to represent ancestors when recombination occurs. An example is shown in Figure 1. By convention at recombination the genes represented by the left upper edge and the lower edge have the same type at the A locus, and similarly for the right edge and the B locus. Probabilistically the graph is described by choosing two edges at random to be joined when a death occurs in the process $\{n(t);\ t \geq 0\}$, and choosing an edge at random to append two edges to when a birth occurs. Mutations occur along the edges of the graph (given the edge lengths) according to a Poisson process of rate $\theta/2 = (\theta_A + \theta_B)/2$, with probabilities θ_A/θ and θ_B/θ of being on locus A or locus B. Knowing the graph and where mutations have occurred along the edges completely determines the configuration of allele pairs in the sample of n genes. Locus A's marginal coalescent tree is found by tracing a path from the n genes upward in the graph, always taking the left path at recombination events, similarly for locus B, taking the right path. The marginal common ancestors may occur in the graph before the (first) common ancestor of the pairs. Figure 1 illustrates the marginal trees in a graph. First marginal ancestors are shown as dots. If $\rho = 0$ there is no recombination and the common ancestors are the same. As $\rho \to \infty$ the waiting time to the common ancestor of the pairs tends to infinity, but the marginal times to common ancestors remain finite, their distribution not depending on ρ.

Hudson (1983), Kaplin and Hudson (1985) use genealogical methods in studying an m-loci model with recombination, where they consider the collection of m correlated family trees.

In a single-locus model the waiting time T_n until there is a common ancestor of n genes is a sum of mutually independent exponential random variables with rates

$n(n-1)/2, \ldots, 1$. The density of T_n is known (see Tavare (1984)) and

$$E(T_n) = \sum_{r=2}^{n} 2/(r(r-1)) = 2(1-n^{-1}).$$

As $n \to \infty$ the distribution of T_n converges to a proper distribution.

Theorem 1. Let W_n be the waiting time until there is a common ancestor of a sample of n genes in the two-locus ancestral graph, then

$$E(W_n) = 2\rho^{-1}\int_0^1 \frac{1-v^{n-1}}{1-v}\,(e^{\rho(1-v)}-1)\,dv. \tag{2.1}$$

Proof. By considering the waiting time until the first event in the birth and death process with rates $\lambda_r = r\rho/2$, $\mu_r = r(r-1)/2$, $r = 1,2,\ldots$, $E(W_1) = 0$, and

$$E(W_n) = \frac{2}{n(n+\rho-1)} + \frac{n-1}{n+\rho-1}E(W_{n-1}) + \frac{\rho}{n+\rho-1}E(W_{n+1}) \tag{2.2}$$

$n = 2,3,\ldots$.

Couple the birth and death process with one where there is a reflecting barrier at $b > 1$ by deleting all excursions $\geq b$. Denote W_n^b as the waiting time to 1 in the modified process. $E(W_n^b)$ satisfies a similar recursion to above, but for $n =$ 2,3, ..., b - 1, and $E(W_b^b) = E(W_{b-1}^b)$. It is straightforward to solve the modified system of equations and obtain

$$E(W_n^b) = 2\sum_{k=2}^{n}(k-2)!\sum_{j=0}^{b-1-k}\rho^j/(j+k)!$$

$n = 2, \ldots, b - 1$.

The limit as $b \to \infty$ is

$$E(W_n) = 2\sum_{k=2}^{n}\sum_{j=0}^{\infty}\rho^j\frac{\Gamma(k-1)\Gamma(j+2)}{\Gamma(j+k+1)\,(j+1)!}$$

$$= 2\rho^{-1}\int_0^1\frac{1-v^{n-1}}{1-v}\left(e^{\rho(1-v)}-1\right)dv. \quad \square$$

Extreme cases of (2.1) are $E(W_n) = 2(1 - n^{-1})$, when $\rho \to 0$ and $E(W_n) \to \infty$ when $\rho \to \infty$. A limit formula is

$$E(W_\infty) = 2\rho^{-1}\int_0^1\frac{e^{\rho u}-1}{u}du.$$

The waiting time W_n can be decomposed into the waiting times spent in the states 2,3, ..., before the process is absorbed at 1.

Theorem 2. Let $W_{n,j}$ be the waiting time spent while there are j ancestors of a sample of n genes in the two-locus ancestral graph, then $W_n = \Sigma_{j=2}^{\infty}W_{n,j}$, and

$$E(W_{n,j}) = 2\sum_{k=2}^{\min(j,n)}(k-2)!\,\rho^{j-k}/j!, \tag{2.3}$$

$j = 2,3,\ldots$

Proof. This is similar to the proof of Theorem 1. $E(W_{n,j})$ satisfies a similar system of equations to (2.2), but with the first term on the right multiplied by the Kronecker delta $\delta_{n,j}$. □

If $\rho = 0$, then $E(W_{n,j}) = 2/(j(j-1))$, agreeing with known results, since then $W_{n,j}$ has an exponential distribution. $E(W_{n,2}) = 1$, not depending on ρ. As $\rho \to \infty$, $E(W_{n,j}) \approx 2\rho^{j-2}/(j(j-1))$. If $n \to \infty$, the upper summation limit in (2.3) is replaced by j.

It is of interest to consider the maximum number of ancestors of a sample that there can be at any time instant before a common ancestor in the two-locus ancestral graph.

Theorem 3. Let M_n be the maximum number of edges in a cross section of a two-locus ancestral graph of a sample of n genes, then

$$P(M_n \le k) = \frac{\Sigma_{j=n-1}^{k-1} j!/\rho^j}{\Sigma_{j=0}^{k-1} j!/\rho^j}, \tag{2.4}$$

$k = n, n+1, \ldots$.

Proof. Denote $p_n(k) = P(M_n \le k)$. Clearly for $k = 1, 2 \ldots$,

$$p_n(k) = \frac{n-1}{n+\rho-1} p_{n-1}(k) + \frac{\rho}{n+\rho-1} p_{n+1}(k),$$

$n = 2, \ldots, k$, and $p_1(k) = 1$, $p_n(k) = 0$ if $n > k$. Denote $u_n(k) = p_n(k) - p_{n-1}(k)$, then

$$\rho u_{n+1}(k) = (n-1)\, u_n(k),$$

$$u_{j+1}(k) = \frac{(j-1)!}{\rho^{j-1}} u_2(k), \quad j = 1, \ldots, k.$$

Summing this equation and using $p_{k+1}(k) = 0$,

$$p_{n+1}(k) - 1 = (p_2(k) - 1) \sum_{j=0}^{n-1} j!/\rho^j, \quad p_2(k) = \frac{\Sigma_{j=1}^{k-1} j!/\rho^j}{\Sigma_{j=0}^{k-1} j!/\rho^j} \tag{2.5}$$

Simplifying (2.5) gives (2.4). □

As $n \to \infty$, $n^{-1}M_n \to 1$, in probability, and from (2.4), $P(M_n > n + d) \approx (n^{-1}\rho)^{d+1}$, $d = 0, 1, \ldots$.

The two-locus ancestral graph will now be examined in greater detail. Given a two-locus graph G construct two marginal coalescent trees T_A and T_B. It is convenient to consider the marginal trees up to the common ancestor vertex in G, although the common ancestor vertices in T_A and T_B may occur earlier. Denote the edge set of a graph by $E(\cdot)$. Partition the edges of G into $A = E(G) \cap E(T_A) \cap E(T_B)^c$, $B = E(G)E(T_A)^C \cap E(T_B)$,

$C = E(G) \cap E(T_A) \cap E(T_B)$, $D = E(G) \cap E(T_A)^C \cap E(T_B)^C$.

. As an example all four types of edge occur in Figure 1. Let E_t be the edges of a cross section of G taken at time t backwards. Denote $n_A(t) = |E_t \cap A|$, $n_B(t) = |E_t \cap B|$, $n_c(t) = |E_t \cap C|$, $n_D(t) = |E_t \cap D|$ and $n(t) = |E_t|$.

Ancestors represented by edges in A do not contain genetic material which has any influence on locus B, similarly for B and locus A, and those represented by D do not contain material influencing either locus A or B.

Our interest is in the Markov process $(n_A(t), n_B(t), n_C(t), n_D(t))$. Think of edges in G as being particles of possible types (1,0), (0,1), (1,1) or (0,0), according as to whether they belong to A, B, C or D. That is, the type of an edge $e \in G$ is $(I_{E(T_A)}(e),\ I_{E(T_B)}(e))$, where $I(\cdot)$ denotes an indicator function. Two edges of types (α, β), (γ, δ) $((\alpha, \beta, \gamma, \delta) \in \{0,1\})$, coalesce to $(\alpha \vee \beta, \gamma \vee \delta)$ ($\vee$ denotes logical or). If recombination occurs to an edge of type (α, β) the edge splits into two edges of types $(\alpha, 0)$ and $(0, \beta)$. Once convinced that this particle evolution is consistent with the type definitions in the partition of G it is clear that $(n_A(t), n_B(t), n_C(t), n_D(t))$ is Markovian.

Transition rates are :

$$
(a, b, c, d) \rightarrow \begin{cases}
(a+1, b+1, c-1, d), & c\rho/2 \\
(a-1, b-1, c+1, d), & ab \\
(a-1, b, c, d), & ac + a(a-1)/2 \\
(a, b-1, c, d), & bc + b(b-1)/2 \\
(a, b, c-1, d), & c(c-1)/2 \\
(a, b, c, d+1), & (a+b+d)\rho/2 \\
(a, b, c, d-1), & d(a+b+c) + d(d-1)/2
\end{cases}
$$

with $a + b + c + d > 1$.

The marginal transition rates of $(n_A(t), n_B(t), n_C(t))$ do not depend on $n_D(t)$, so it too is Markovian.

Transition rates are :

$$(a,b,c) \to \begin{cases} (a+1, b+1, c-1), & c\rho/2 \\ (a-1, b-1, c+1), & ab \\ (a-1, b, c), & ac + a(a-1)/2 \\ (a, b-1, c), & bc + b(b-1)/2 \\ (a, b, c-1), & c(c-1)/2, \end{cases}$$

with $a + b + c > 1$.

This process is used in Ethier and Griffiths (1990a), with a different interpretation, in showing that the two-locus measure-valued diffusion is ergodic. Although $n(t)$ is not a monotonic process, both $n_A(t) + n_C(t)$ and $n_B(t) + n_C(t)$ are non-increasing processes, which are eventually absorbed into 1,1.

It is clear how to extend the two-locus graph to multiple loci, although the extension is not studied here. Consider m loci with a total recombination rate ρ. Given a recombination event, it splits a gene between locus j and $j + 1$ with probability ρ_j/ρ, $j = 1, \ldots, m-1, \rho = \rho_1 + \ldots + \rho_{m-1}$. The ancestral graph is still similar to the illustration in Figure 1. Edges are labeled by elements of $\{0, 1\}^m$. Coalescence is similar to the two-locus case, and if recombination occurs between locus j and $j + 1$ to an edge of type $(\alpha_1, \ldots, \alpha_m)$ the upper joining left and right edges have respective types $(\alpha_1, \ldots, \alpha_j, 0, \ldots, 0)$, $(0, \ldots, 0, \alpha_{j+1}, \ldots, \alpha_m)$. Results of Theorems 1,2,3 apply also to multiple loci.

Hudson(1983), Kaplin and Hudson (1985) use the process (2.6) and its extension to m-loci in their study of the homozygosity and number of segregating sites in a sample of genes. They were not concerned with the total graph, but just the correlated trees at the m loci.

A system of equations which is useful in several theorems is the following. Let υ and β be functions on $\mathbf{N} \times \mathbf{N} \times \mathbf{N} \times \mathbf{R}$ satisfying

$$\begin{aligned} \nu(a,b,c;\rho) &= \beta(a,b,c;\rho) + \frac{c\rho}{n(n-1)+c\rho}\upsilon(a+1, b+1, c-1;\rho) \\ &+ \frac{2ab}{n(n-1)+c\rho}\nu(a-1\ \ b-1, c+1;\rho) + \frac{a(a+2c-1)}{n(n-1)+c\rho}\nu(a-1,b,c;\rho) \\ &+ \frac{b(b+2c-1)}{n(n-1)+c\rho}\nu(a,b-1,c;\rho) + \frac{c(c-1)}{n(n-1)+c\rho}\nu(a,b,c-1;\rho). \end{aligned} \tag{2.7}$$

where $n = a + b + c$, $a + c > 1$, $b + c > 1$. The function β and boundary conditions $\nu(1, 1, 0;\rho)$, $\nu(0, 0, 1; \rho)$, $\nu(1, 0, 0; \rho)$, $\nu(0, 1, 0; \rho)$ are assumed to be known.

The system arises by considering the first transition in the process $(n_A(t), n_B(t), n_C(t))$ beginning at (a,b,c). The overall transition rate away from (a,b,c) is $(n(n - 1) + c\rho)/2$. Proofs of Theorems 4,6,7 are straightforward using systems of equations of the form (2.7), so are omitted. Ethier and Griffiths (1990b) show how to solve systems of equations such as (2.7) numerically. Defining the degree of $\nu(a, b, c;\rho)$ as $a + b + 2c$, terms on the right side of (2.7) are of degree less than or equal to $a + b + 2c$. Knowing terms of lesser degree the others can be found by solving a tridiagonal system of equations.

Let $W_{A,n}$ and $W_{B,n}$ be the respective waiting times until a sample of n genes has common ancestors at locus A and locus B.

Theorem 4. Let $\xi(a, b, c;\rho)$ be the expected waiting time until $n_A(t) + n_C(t) = 1$, $n_B(t) + n_C(t) = 1$, beginning with $n_A(0) = a$, $n_B(0) = b$, $n_C(0) = c$.

$E(\max(W_{A,n}, W_{B,b})) = \xi(0, 0, n; \rho)$ is the expected waiting time until a sample of n genes has common ancestors at locus A and locus B. ξ satisfies the system (2.7) with $\beta(a, b, c; \rho) = 2/(n(n - 1) + c\rho)$ and $\xi(1, 1, 0; \rho) = 0$, $\xi(0, 0, 1; \rho) = 0$, $\xi(1, 0, 0; \rho) = 0$, $\xi(0, 1, 0; \rho) = 0$. □

Illustrative values of the waiting time are shown in Table 1 for a sample of size 50.

Table 1

Expected waiting time to marginal common ancestors, n = 50.

ρ	0	0.1	0.5	1.0	2.0	5.0	10.0	20.0	∞
μ	1.96	2.01	2.14	2.25	2.36	2.46	2.50	2.51	2.52

The waiting time until a sample has marginal common ancestors is finite as $\rho \to \infty$, being bounded below by $2(1 - n^{-1})$ and above by $4(1 - n^{-1})$, however the waiting time until a common ancestor in G tends to infinity. $\xi(a, b, c; \rho)$ is monotonic in ρ and therefore bounded by $\xi(a+c, b+c, 0; \infty)$. One argues that if $\rho = \infty$, edge types (1,1) are instantaneously split into a pair (1,0) and (0,1), so $\xi(a, b, c; \infty) = \xi(a+c, b+c, 0; \infty)$. In this case there is an easy recursion,

$$\xi(a,b,0;\infty) = \frac{2}{n(n-1)} + \frac{2ab}{n(n-1)}\xi(a,b,0;\infty)$$

$$+\frac{a(a-1)}{n(n-1)}\xi(a-1,b,0;\infty) + \frac{b(b-1)}{n(n-1)}\xi(a,b-1,0;\infty),$$

$a, b = 1,2, \ldots, n = a+b > 1$ and $\xi(1,1,0;\infty) = 0$.

Simplifying,

$$\xi(a,b,0;\infty) = \frac{2}{a(a-1)+b(b-1)} + \frac{a(a-1)}{a(a-1)+b(b-1)}\xi(a-1,b,0;\infty)$$

$$+\frac{b(b-1)}{a(a-1)+b(b-1)}\xi(a,b-1,0;\infty), \tag{2.8}$$

$a, b = 2,3, \ldots$ and

$$\xi(a,1,0;\infty) = 2(1-a^{-1}),\ a = 1,2,\ldots,\quad \xi(1,b,0;\infty) = 2(1-b^{-1}),\ b = 1,2,\ldots\ .$$

If $\rho = \infty$, $W_{A,n}$ and $W_{B,n}$ are independent and their distribution known. It is therefore possible to work out a (series) formula for $\xi(a,b,0;\infty)$, but (2.8) is convenient for calculation. Illustrative values of the waiting time when $\rho = \infty$ are shown in Table 2.

Table 2

Expected waiting time to marginal common ancestors, $\rho = \infty$.

n	2	10	20	50	100	500	1000
μ	1.50	2.36	2.46	2.52	2.54	2.56	2.56

If $\rho = 0$ the process essentially behaves as a single-locus one and $\xi(0,0,n;0) = 2(1-n^{-1})$.

$E(\min(W_{A,n}, W_{B,n}))$ can be found from the relationship

$$E(\min(W_{A,n},W_{B,n}) + \max(W_{A,n},W_{B,n})) = E(W_{A,n}+W_{B,n}) = 4(1-n^{-1}).$$

It is possible to find the distribution of $\max(W_{A,2}, W_{B,2})$ and $\min(W_{A,2}, W_{B,2})$ explicitly.

Theorem 5. The joint Laplace transform

$$\psi(\lambda_1,\lambda_2) = E(\exp\{-\lambda_1\max(W_{A,2},W_{B,2}) - \lambda_2\min(W_{A,2},W_{B,2})\}) =$$

$$\left(1+\frac{\lambda}{1+\rho}\right)^{-1}\left(\frac{1}{1+\rho}+\frac{\rho}{1+\rho}\phi(\lambda_1,\lambda_2)\right),$$

where

$$\phi(\lambda_1,\lambda_2)=(1-\beta(\phi_1(\lambda_1,\lambda_2)-1))^{-1}\,\phi_2(\lambda_1,\lambda_2),\ \lambda=\lambda_1+\lambda_2,\ \beta=\frac{2\rho(4+\rho)}{\rho^2+13\rho+18},$$

and

$$\phi_1(\lambda_1,\lambda_2) = \left(1+\frac{2\lambda}{6+\rho}\right)^{-1}\left(\gamma\left(1+\frac{\lambda}{6}\right)^{-1}+(1-\gamma)\left(1+\frac{\lambda}{1+\rho}\right)^{-1}\right),\ \gamma=\frac{1+\rho}{4+\rho},$$

$$\phi_2(\lambda_1,\lambda_2) = \left(1+\frac{2\lambda}{6+\rho}\right)^{-1}\left(\delta_1(1+\lambda_1)^{-1}\left(1+\frac{\lambda}{6}\right)^{-1}+\delta_2\left(1+\frac{\lambda}{1+\rho}\right)^{-1}+\delta_3(1+\lambda_1)^{-1}\right),$$

$$\delta_1=\frac{\rho(1+\rho)}{\rho^2+13\rho+18},\delta_2=\frac{6}{\rho^2+13\rho+18},\delta_3=\frac{12(1+\rho)}{\rho^2+13\rho+18}.$$

An alternative form for the Laplace transform is

$$\psi(\lambda_1,\lambda_2) = \frac{A}{(1+\lambda_1)B} \tag{2.9}$$

where

$$A = \lambda_1(2\lambda^2+\lambda(\rho+18)+2\rho+36)+2\lambda^2+\lambda(5\rho+18)+2\rho^2+26\rho+36$$

$$B = 2\lambda^3+\lambda^2(3\rho+20)+\lambda(\rho^2+19\rho+54)+2\rho^2+26\rho+36.$$

The mean values are

$$E(\max(W_{A,2},W_{B,2})) = \frac{3\rho^2+38\rho+36}{2(\rho^2+13\rho+18)},$$

$$E(\min(W_{A,2},W_{B,2})) = \frac{\rho^2+14\rho+36}{2(\rho^2+13\rho+18)}. \tag{2.10}$$

Proof. Let $f_{abc}(\lambda_1,\lambda_2)$ be the Laplace transform of the waiting times, given $n_A(0)=a$, $n_B(0)=b$, $n_C(0)=c$. The waiting times in state (a,b,c) have a Laplace transform

$$g_{abc}(\lambda_1,\lambda_2) = \left(1+\frac{2\zeta}{n(n-1)+c\rho}\right)^{-1},$$

where

$$\zeta = \begin{cases} \lambda \text{ if } a+b>1, b+c>1 \\ \lambda_1 \text{ if } a+c=1, b+c>1 \text{ or } a+c>1, b+c=1 \end{cases}$$

Let $f_{abc} \equiv f_{abc}(\lambda_1, \lambda_2)$ and $g_{abc} \equiv g_{abc}(\lambda_1, \lambda_2)$. By considering the first transition from particular values of (a,b,c),

$$f_{002} = g_{002}\left(\frac{\rho}{1+\rho}f_{111} + \frac{1}{1+\rho}\right),$$

$$f_{111} = g_{111}\left(\frac{\rho}{6+\rho}f_{220} + \frac{2}{6+\rho}f_{002} + \frac{4}{6+\rho}f_{101}\right),$$

$$f_{101} = g_{101}\left(\frac{\rho}{2+\rho}f_{210} + \frac{2}{2+\rho}\right),$$

$$f_{210} = g_{210}\left(\frac{2}{3}f_{101} + \frac{1}{3}\right),$$

$$f_{220} = g_{220}\left(\frac{2}{3}f_{111} + \frac{1}{3}f_{210}\right).$$

The required Laplace transform is f_{002}. Solving the equations,

$$f_{101} = f_{210} = (1+\lambda_1)^{-1} \text{and} \quad f_{111} = \phi(\lambda_1, \lambda_2),$$

Substituting in the first equation produces the Laplace transform in the statement of the theorem. It is straightforward to find the means (2.10) from the Laplace transform (2.9). □

Corollary.

$$E(exp\{-\lambda_1 W_{A,2} - \lambda_2 W_{B,2}\}) = \frac{1}{2}(\psi(\lambda_1, \lambda_2) + \psi(\lambda_2, \lambda_1)).$$

Proof. This is clear by symmetry. □

The first form of $\psi(\lambda_1, \lambda_2)$ in Theorem 5 is an expression in terms of a mixture. Another way to invert $\psi(\lambda_1, \lambda_2)$ is to use a partial fraction expansion of (2.9). The distribution has a singular component where $W_{A,2} = W_{B,2}$. Express (2.9) as

$$\psi(\lambda_1, \lambda_2) = \psi_0(\lambda) + \psi_1(\lambda_1, \lambda_2)$$

where

$$\psi_0(\lambda) = (2\lambda^2 + (\rho+18)\lambda + 2\rho + 36)/B$$

$$\psi_1(\lambda_1, \lambda_2) = (1+\lambda_1)^{-1} 2\rho\,(2\lambda+\rho+12)/B. \tag{2.11}$$

The singular part of the distribution has Laplace transform $\psi_0(\lambda)$ and the continuous part $\psi_1(\lambda_1, \lambda_2)$. In particular

$$P(W_{A,2} = W_{B,2}) = \psi_0(0) = \frac{\rho+18}{\rho^2+13\rho+18}. \tag{2.12}$$

It is possible to show that $B(\lambda, \rho)$, the cubic in λ appearing in the denominator of $\psi(\lambda_1, \lambda_2)$, has three distinct negative roots for all $\rho \geq 0$. The Laplace transforms $\psi_0(\lambda)$, $\psi_1(\lambda_1, \lambda_2)$, $\psi(\lambda, 0)$, $\psi(0, \lambda)$, can be inverted by partial fraction techniques after finding the roots of $B(\lambda, \rho)$ for a given numerical value of ρ. An expression of the form

$$(1+\lambda_1)^{-1}(1+\lambda/\beta)^{-1}, \beta > 0$$

occurring in the expansion of $\psi(\lambda_1, \lambda_2)$ corresponds to a density

$$\beta e^{-u-(\beta-1)v}, 0 < v < u < \infty.$$

If $\rho = 0$ both the waiting times are identical, and have an exponential distribution with mean 1. If $\rho \to \infty$ the waiting times are the maximum and minimum of two independent exponential random variables, with mean 1, and so

$$P(\max(W_{A,2}, W_{B,2}) \leq w) = (1-e^{-w})^2, \quad P(\min(W_{A,2}, W_{B,2}) \leq w) = 1-e^{-2w}.$$

Hudson and Kaplan (1985) introduced the idea of the number of recombination events in a sample of genes in an m-loci model with recombination. With two loci their definition of the number of recombination events occurring in a sample of n genes is the number of events occurring to ancestors of the sample which have both loci belonging to the marginal trees. Denote their number of recombination events by $\mathring{R}_n$. Ethier and Griffiths (1990a) define their number of recombination events R_n as the number of recombination events in a two-locus graph before the common ancestor. Of course $R_n \geq \mathring{R}_n$. They show that R_n is distributed as the number of steps to the right in a random walk starting at n with absorbing state 1 and transition probabilities for $m \geq 2$,

$$m \to \begin{cases} m-1, & \text{probability } (m-1) \quad (m-1+\rho), \\ m+1, & \text{probability } \rho \quad (m-1+\rho) \end{cases}.$$

An explicit formula for the probability generating function of R_n is given in Ethier and Griffiths (1990b), where it is also shown that

$$E(R_n) = \rho\int_0^1 \frac{1-(1-v)^{n-1}}{v} e^{\rho v} dv.$$

They also show that there is a stochastic representation related to the random walk which gives the two-locus sampling distribution. This is a particularly attractive representation because of its simplicity, and is a good reason for having such an extended definition of ancestors. Here is a description of the process. Run a random walk starting at n until it hits 1.

Transition probabilities of the random walk are :

$$r \rightarrow \begin{cases} r-1, & (r-1)\ (r-1+\theta+\rho), \\ r, & \theta\ (r-1+\theta+\rho), \\ r+1, & \rho\ (r-1+\theta+\rho), \end{cases}$$

where $\theta = \theta_A + \theta_B$.

Let τ be the hitting time of 1. Construct samples of genes (pairs of loci) at $\tau - 1, \tau - 2, \ldots, 0$ in the following way. Start with a sample of size 1 at τ. Let $\{U_t, t = 0, 1, 2, \ldots\}$ be the random walk and $\{Z_t = U_{t+1} - U_t, t = 0, 1, \ldots\}$. If $Z_t = -1$, duplicate a sample member at random. If $Z_t = 0$, choose a sample member to mutate, at either locus A or B with respective probabilities θ_A/θ or θ_B/θ. If $Z_t = +1$, choose two members of the sample without replacement to recombine. The sample of n at time 0 has the two-locus sampling distribution.

Theorem 6. Let $\eta(a, b, c; \rho)$ be the expected number of transitions $(n_A(t), n_B(t), n_C(t)) \rightarrow (n_A(t^+) + 1, n_B(t^+) + 1, n_c(t^+) - 1)$ before $n_A(t) + n_C(t) = 1$, $n_B(t) + n_C(t) = 1$ beginning with $n_A(0) = a$, $n_B(0) = b$, $n_C(0) = c$.

Then $E(\mathring{R}_n) = \eta(0, 0, n; \rho)$. η satisfies the system (2.7) with $\beta(a, b, c; \rho) = c\rho/(n(n - 1) + c\rho)$ and $\eta(1, 1, 0; \rho) = 0$, $\eta(0, 0, 1; \rho) = 0$, $\eta(1, 0, 0;\rho) = 0$, $\eta(0, 1, 0; \rho) = 0$. A particular case is $E(\mathring{R}_2) = 6\rho/(6 + \rho)$. □

Table 3 illustrates results from Theorem 6 for a sample of size 50.

Table 3

Recombination events in a samples' ancestors.

ρ	$E(R_{50})$	$E(\mathring{R}_{50})$
0	0	0
0.1	4.58×10^{-1}	4.43×10^{-1}
0.5	2.52	2.13
1.0	5.78	4.07
2.0	1.62×10	7.51
5.0	2.12×10^{2}	1.57×10
10.0	2.49×10^{4}	2.56×10
20.0	5.12×10^{8}	3.94×10

As $\rho \to \infty$, $R_n \to \infty$, but $\mathring{R}_n < \infty$ and has a proper limit distribution. Using an argument similar to that for $\xi(a, b, c; \infty)$, $\eta(a, b, c; \infty) = c + \eta(a + c, b + c, 0; \infty)$. $\eta(a, b, c; \infty)$ satisfies a similar system of equations to (2.8), but with the term $2/(a(a-1) + b(b-1))$ replaced by $2ab/(a(a-1) + b(b-1))$, $\eta(a, 1, 0; \infty) = 2\Sigma_{j=1}^{a-1} j^{-1}$, a = 2,3, ..., $\eta(1, b, 0; \infty) = 2\Sigma_{j=1}^{b-1} j^{-1}$, $b = 2,3, \ldots$, , and

$\eta(1, 1, 0; \infty) = 0$. Thus if $\rho = \infty$, $E(\mathring{R}_n) = n + \eta(n, n, 0; \infty)$.

If the first time that $n_A(t) + n_C(t) = 1$, $n_B(t) + n_C(t) = 1$ is when $n_A(t) = 0$, $n_B(t) = 0$, $n_C(t) = 1$ then the first common ancestor in the two-locus graph and the two first marginal common ancestors coincide. This produces the following theorem.

Theorem 7. Let

$$T = \inf\{t \geq 0;\ n_A(t) + n_C(t) = 1,\ n_B(t) + n_C(t) = 1\}.$$

and

$$\zeta(a, b, c; \rho) = P(n_C(T) = 1 \mid n_A(0) = a, n_B(0) = b, n_C(0) = c).$$

Then ζ satisfies the system (2.7) with $\beta(a, b, c; \rho) = 0$ and $\zeta(1, 1, 0; \rho) = 0$, $\zeta(0, 0, 1; \rho) = 1$, $\zeta(1,0, 0; \rho) = 0$, $\zeta(0, 1, 0; \rho) = 0$. A particular case is

$$\zeta(0,0,2;\rho) = \frac{(2+\rho)(9+13\rho)}{(1+\rho)(\rho^2+17\rho+18)}. \quad \square$$

Table 4 illustrates probabilities from Theorem 7 for $n = 50$.

Table 4

Probability that marginal and two-locus ancestors are identical, $n = 50$.

ρ	0	0.1	0.5	1.0	2.0	5.0	10.0	20.0
Probability	1.0	0.998	0.969	0.909	0.785	0.525	0.326	0.182

It is possible to modify coalescent trees so that ancestral lines are lost by coalescence or by mutation (Griffiths (1980), Tavaré (1984)). With a single locus the number of ancestral lines at time t back is a death process with rate $n(n + \theta - 1)/2$. Lines are lost by coalescence, at rate $n(n - 1)/2$ and by mutation at rate $n\theta/2$. In the two-locus case the process $(n_A(t), n_B(t), n_C(t))$ can be modified to include loss by mutation.

Transitions rates are:

$$(a,b,c) \to \begin{cases} (a+1, b+1, c-1), & c\rho/2 \\ (a-1, b-1, c+1), & ab \\ (a-1, b, c), & ac + a(a-1)/2 + a\theta_A/2 \\ (a, b-1, c), & bc + b(b-1)/2 + b\theta_B/2 \\ (a, b, c-1), & c(c-1)/2 \\ (a+1, b, c-1), & c\theta_B/2 \\ (a, b+1, c-1), & c\theta_A/2, \end{cases}$$

with $a + b + c > 1$.

The expected waiting time until $n_A(t) + n_C(t) = 1$ and $n_B(t) + n_C(t) = 1$ in Theorem 4 can be modified by considering the transitions (2.13).

One way to derive the probability generating function, $P_n(z)$, for the number of allele types from a sample of n from the one-locus model is to consider back in time whether lines are lost by mutation or coalescence. Each line lost by mutation corresponds to a distinct allele type in the sample. By considering the first event back in time, for $n \geq 2$,

$$(n+\theta-1)P_n(z) = (n-1)P_{n-1}(z) + \theta z P_n(z). \tag{2.13}$$

Solving (2.13) with the boundary condition $P_1(z) = z$,

$$P_n(z) = \prod_{j=1}^{n} \frac{\theta z + j - 1}{\theta + j - 1}.$$

The proof of the next theorem, in the two-locus case, is clear using this approach. The theorem is also proved in Ethier and Griffiths (1990b) by applying the generator of the measure-valued diffusion. It is possible to compute the joint distribution of the number of allele types at the two loci, but there is no simple formula for it.

Theorem 8. Let (K, L) be the number of allele types at locus A and locus B in a sample of n genes. The *pgf* of (K, L) is $P(0, 0, n; z_A, z_B)$ in the system of equations

$$(n(n-1) + c\rho + (a+c)\theta_A + (b+c)\theta_B) P(a, b, c; z_A, z_B) =$$

$$c\rho P(a+1, b+1, c-1; z_A, z_B) + 2abP(a-1, b-1, c+1; z_A, z_B)$$

$$+ a(a + 2c - 1 + \theta_A z_A) P(a-1, b, c; z_A, z_B)$$

$$+ b(b + 2c - 1 + \theta_B z_B) P(a, b-1, c; z_A, z_B)$$

$$+ c(c-1) P(a, b, c-1; z_A, z_B)$$

$$+ c\theta_B z_B P(a+1, b, c-1; z_A, z_B) \quad + c\theta_A z_A P(a, b+1, c-1; z_A, z_B).$$

Boundary conditions are $P(1, 1, 0; z_A, z_B) = P(0, 0, 1; z_A, z_B) = z_A z_B$. □

References

[1] Ethier, S. N. and Griffiths, R. C. (1990a). The neutral two-locus diffusion model as a measure-valued diffusion. *Adv. Appl. Prob.* **22**, 773-786.

[2] Ethier, S. N. and Griffiths, R. C. (1990b). On the two-locus sampling distribution. *J. Math. Biol.* To appear.

[3] Griffiths, R. C. (1980). Lines of descent in the diffusion approximation of neutral Wright-Fisher models. *Theor. Pop. Biol.* **17** 37-50.

[4] Hudson, R. R. (1983). Properties of a neutral allele model with intragenic recombination. *Theor. Pop. Biol.* **23** 183-201.

[5] Hudson, R. R. (1985). The sampling distribution of linkage disequilibrium under an infinite allele model without selection. *Genetics* **109** 611-631.

[6] Hudson, R. R. and Kaplan, N. L. (1985). Statistical properties of the number of recombination events in the history of a sample of DNA sequences. *Genetics* **111** 147-164.

[7] Kaplin, N. L. and Hudson, R. R. (1985). The use of sample genealogies for studying a selectively neutral m--loci model with recombination. *Theor. Pop. Biol.* **28** 382-396.

[8] Kingman, J. F. C. (1982). The coalescent. *Stochastic Process. Appl.* **13** 235-248.

[9] Tavaré, S. (1984). Line-of-descent and genealogical processes, and their applications in population genetics models. *Theor. Pop. Biol.* **26** 119-164.

WHEN DID JOE'S GREAT . . . GRANDFATHER LIVE? OR: ON THE TIME SCALE OF EVOLUTION.

Peter Jagers, Olle Nerman, and Ziad Taib
Chalmers University of Technology and Gothenburg University

Abstract

In a general supercritical branching process the distribution is found of the time back to the birth of the *n*:th father of a random individual. For single-type processes a probability of mutation (at a site of a specified gene) is introduced. With the infinite alleles interpretation that every mutation occurring is unique the class of individuals with the same allele can be viewed as a generalized "macroindividual". Thereby the question of when the first of a series of *n* mutations occurred is reduced to finding the birth time of the *n*:th father in the process of macroindividuals. This is done.

1. Introduction The three great problems of population dynamics are the extinction, the growth, and the composition of populations. The first makes sense already in the simplest of branching processes, the Galton-Watson model, which was indeed born out of it. The second requires a physical time, not only generation counting, and for the third problem one should have access also to the birth order of siblings and times between their births, and preferably a possibility to discern between individuals of different inherited types.

This leads to general, multi-type branching processes (Jagers and Nerman 1984 (the one-type case), Nerman 1984, Jagers 1989 (abstract type spaces)). When an individual is born it inherits a *type*, s, from a *type space*, S with a σ-algebra S. The type, which you may think of as a genotype, determines a probability kernel, $P(s,\cdot)$, over a *life space*, (Ω, A), of all possible life careers. The life space may be very rich in order to contain all relevant aspects of individual life, and is best thought of as an abstract measurable space. What has to be defined on it - otherwise there would be no population to talk of - is a *reproduction process* for individuals. It will be denoted by ξ and viewed as a point process on $S \times \mathbf{R}_+$, the first coordinate of any point yielding the type of that child, the second the father's age at the birth[1]. The processes is supposed finite on bounded subsets of $\mathbf{R}_+$, so that the children can be numbered in birth order (arbitrarily for twins). If σ_k, τ_k are thus the type of the k:th child and father's age at his birth, then

$$\xi(A \times B) = \#\{k; \sigma_k \in A, \tau_k \in B\}, A \in A, B \in B,$$

B denoting the Borel algebra on $\mathbf{R}_+$. A biologically basic, though not mathematically fundamental, entity defined on the life space is the *life span* ,

$$\lambda : \Omega \to \bar{\mathbf{R}}_+ \quad ,$$

the extended non-negative half-line.

The individuals to lead these lives are denoted in the classical Ulam-Harris fashion: the ancestor is 0, the first generation is $\{1,2,...\} = N$, the n:th is $N^n, N^0 = \{0\}$, and the set of all possible individuals is written

$$I = \bigcup_{n \geq 0} N^n.$$

If individuals and sets of individuals are partially ordered by descent (x stems from y if $y = 0$ or y is the first part of x, A stems from $B \subset I$ if for all $x \in A$ there is some $y \in B$ such that x stems from y), then the *life kernel* $P(s,\cdot)$, $s \in S$ determines a Markov field over sets of individuals (Jagers 1989) and the flow of time via the following dynamics:

- The ancestor 0 is born at time 0 and of an arbitrary type $\sigma_0 = s \in S$.

- He gets $\xi_0(S \times \mathbf{R}_+)$ children, *viz.* 1,2... $\xi_0(S \times \mathbf{R}_+)$, at times $\tau_1, \tau_2 ... \tau_{\xi_0(S \times \mathbf{R}_+)}$ and they are of types $\sigma_1, \sigma_2 ... \sigma_{\xi_0(S \times \mathbf{R}_+)}$. These are measurable functions of the ancestor's life, chosen from the life space Ω according to $P(\sigma_0, \cdot)$.

- The individual $x = (x_1, x_2, ... x_n)$ (= $x_1x_2...x_n$ for short), $x_i \in N$, is the x_n:th child of $x_1x_2...x_{n-1}$. He is of type σ_x and is born at time τ_x which is the sum of 0's age τ_{x_1} at begetting x_1 , the latter's age at giving birth to his x_2:th child *etc.*. The interpretation of $\tau_x = \infty$ is, of course, that x is never born.

The probability measure thus arising on the *population space* $(S \times \Omega^I, S \times A^I)$ will be denoted by $\mathbf{P}_s$ for $\sigma_0 = s \in S$ with expectation operator $\mathbf{E}_s$. If the the starting type was chosen according to some distribution, say π, over S. We write $\mathbf{P}_\pi$ and $\mathbf{E}_\pi$.

A couple of rather technical assumptions (*cf.* Jagers 1989) are needed. They can be summarized by saying that the process should be

- Malthusian, *i.e.* there is a Malthusian growth parameter α,

- supercritical, *i.e.* $0 < \alpha < \infty$,

- irreducible, *i.e.* individuals of almost all (with respect to a *stable type distribution* π, see beneath) types can have descendants of almost all types,

- strongly α-recurrent, meaning that the mean age at childbearing in the population is (positive and) finite, and

- spread-out in the sense that the *reproduction kernel*

$$\mu(s, A \times B) = \mathbf{E}_S[\xi(A \times B)] = \int_\Omega \xi(A \times B)(\omega) P(s, d\omega)$$

is so with respect to the product of the stable type and Lebesgue measures.

The strict definition of the stable type measure π is that it is the maximal measure satisfying

$$\int_0^\infty e^{-\alpha t} \mu(\pi, A \times dt) = \pi(A), \mu(\pi, A \times B) = \int_S \mu(s, A \times B) \pi(ds),$$

It is assumed normed to a probability measure.

Now assume that there is given a so called *random characteristic i.e.* a weight, that for each individual might be influenced by his type, age, life and progeny, but not of his ancestors' lives. (Jagers 1975 gives many examples, one will follow here.) Make the dependence upon age explicit, writing the characteristic $\chi(a)$, a for age. At time t sample an individual at random from among all those born (*the total population*) or from among all those alive at that time (or possibly from some other subset of the total population). Call that individual *Joe* and check his χ-value. The population development given, its expectation will be the average χ-value in the population now at time t. Then, let $t \to \infty$. Will this conditional expected χ-value stabilize? Under certain technical conditions (including the famous $x\log x$) the answer is "yes". The limit, which we shall call simply Joe's expected χ-value in the stable case, is given by the following theorem, stated quite loosely here. (For a precise formulation and proof *cf.* Jagers 1989.) In it, and beneath, *AGE* is an exponentially distributed random variable with parameter α independent of everything else, the index π indicates that the expectation is taken with respect to a branching process whose ancestor had a type, random according to the stable type distribution, χ and λ refer to this ancestor, and the hat to Laplace transform, $\hat{\chi}(\alpha) = \alpha\int_0^\infty e^{-\alpha t}\chi(t)\,dt$.

Theorem 1. *In a stable population Joe's expected* χ*-value is* $\mathbf{E}_\pi[\hat{\chi}(\alpha)]$ *if he was sampled from the total population,* $\mathbf{E}_\pi[\hat{\chi}(\alpha) \mid AGE > \lambda]$ *if sampled from among those alive.*

Famous special cases of this theorem are the stable age distribution of demography (the characteristic being the indicator of age intervals) and the mitotic index of cell kinetics (where the characteristic is the indicator of the cell cycle time in mitosis). The theorem is itself a special case of a description of the stable population generally, *i.e.* also for characteristics influenced by Joe's progenitors, cousins *etc.* , *cf.* Jagers and Nerman 1984 for the one-type case and Nerman 1984 for the general theory. Since Nerman's paper is not so easily available we shall, however use the theorem in the next section. For the one-type case *cf.* , however, Nerman and Jagers 1984. But certainly the general stable population provides the proper context for the problem to be treated.

2. When Was Joe's *n*:th Father Born? What is, thus, the probability that Joe's *n*:th father, i.e. Joe's father's father's ... father (*n* times) was born no more than *u* time units ago? To make use of the theorem we must find a characteristic that counts an individual if his *n*:th father satisfies this. Since it may not be influenced by ancestors of the individual it measures, we have to refer it to the *n*:th father himself, giving him the weight *k* if he has n:th generation descendants within time *u*:

$$\chi^u(a) = \sum_{n \in N^n} 1_{\{\tau_x \le a \le u\}} .$$

(In this first step we do not bother whether the *n*:th generation descendant is alive or not.) Clearly

$$\hat{\chi}^u(a) = \sum_{n \in N^n} \alpha \int_{\tau_x}^{u} e^{-\alpha a} da 1_{\{r_x \le u\}} = \sum_{n \in N^n} (e^{-\alpha\tau_x} - e^{-\alpha u}) 1_{\{\tau_x \le u\}} .$$

For functions of a non-negative real variable let * denote convolution, and for functions on $S \times \mathbf{R}_+$ let it stand for transition on S and convolution on $\mathbf{R}_+$. Thus,

$$\mu^{*2}(A \times B) = \mu * \mu(A \times B) = \int_{S \times \mathbf{R}_+} \mu(r, A \times B - t)\mu(s, dr \times dt)$$

Also write

$$\mu_\alpha(s, dr \times dt) = e^{-\alpha t}\mu(s, dr \times dt) .$$

Since by induction

$$\sum_{x \in N^n} P_\pi(\tau_x \le u) = \mu^{*n}(\pi, S \times [0, u]),$$

it follows that

$$\mathrm{E}_\pi[\hat{\chi}^u(\alpha)] = \int_0^u (e^{-\alpha t} - e^{-\alpha u})\, \mu^{*n}(\pi, S \times dt) = E_\alpha * \mu_\alpha^{*n}(\pi, S \times [o, u]),$$

E_α denoting the exponential distribution with parameter α.

The case of proper interest, Joe alive, is somewhat more difficult in this setting. (It is easy if you have the general stable population theory.) We cannot apply the conditional version of the theorem, since the characteristic has been carried back to referring to Joe's n:th father, rendering him the Joe of the theorem. Instead we have to consider a characteristic

$$\chi_l^u(a) = \sum_{n \in N^n} 1_{\{\tau_x \le a \le u, \lambda_x > a - r_x\}},$$

the index x indicating that not only the birth-time refers to the individual x but also the life-span. Calculating as above leads to an expression involving, besides the reproduction function, also the *life span distribution, L* of an individual whose type is determined according to the stable type distribution, π,

$$\mathrm{E}_\pi[\hat{\chi}_l^u(a)] = \mathrm{E}_\pi\left[\sum_{n \in N^n} \alpha \int_{\tau_x}^{u \wedge (\tau_x + \lambda_x)} e^{-\alpha a}\, da\, 1_{(\tau_x \le u\}}\right] = \dots =$$

$$= \int_0^u \left(\int_0^{(u-t)} e^{-\alpha v}(1 - L(v))\, dv\right) \mu^{*n}(\pi, S \times dt).$$

This is, however, not the correct expression: we have to norm by the probability of being alive,

$$\int_0^\infty e^{-\alpha t}(1 - L(t))\, dt.$$

Then it follows that the probability of Joe's n:th father being born no more than time u ago has the form

$$L_\alpha * \mu_\alpha^{*n}(\pi, S \times [0, u]),$$

L_α being the stable age distribution

$$\frac{\int_0^u e^{-\alpha t}(1 - L(t))\, dt}{\int_0^\infty e^{-\alpha t}(1 - L(t))\, dt}.$$

If life length is infinite, this reduces to the exponential distribution, of course.

We specialize now to the one-type case where there is a proper reference for the stable population derivation of this result, Nerman and Jagers 1984. Then $\mu(\pi, S \times [0, u])$ is merely the *reproduction function* $\mu(u)$ and $\mu_\alpha(\pi, S \times [0, u])$ a probability measure μ_α on $\mathbf{R}_+$, the stable distribution of age at childbearing. Its expectation

$$\beta = \int_0^\infty t\mu_\alpha(dt) = \int_0^\infty te^{-\alpha t}\mu(dt)$$

is the *mean age at childbearing*. Its variance will be denoted by v^2.

The central limit theorem thus yields that the time back to Joe's n:th father's birth is, approximately,

$$X + n\beta + v\sqrt{n}N(0, 1),$$

X being exponential (α) if Joe was sampled from the total population and following the stable age distribution, if he was chosen from those alive, and independent of the standard normal variable $N(0,1)$ in both cases. If there is further information about Joe this will reflect itself only on X. *E.g.* if Joe is known to be 65, the time back will be

$$65 + n\beta + v\sqrt{n}N(0, 1).$$

Example. In the last mentioned case we find out when Joe's greatgreatgreatgreatgreatgreatgreatgreatgrandfather was born. Assume that $\beta = 30$ and $v = 7$. An approximately 95% prediction (or, rather, postdiction) interval is then

$$1989 - 65 - 300 \pm 2 \times 7 \times \sqrt{10} \approx 1624 \pm 46 = [1578, 1670].$$

3. The Time Scale of Evolution. Through the advent of molecular biology it has become possible to study the process of evolution at the level of nucleic and amino acid sequences. There is a host of literature, *cf.* Kimura 1979, Wilson *et al.* 1987, or in a branching process context Taib 1987. Much of the statistically oriented work, *cf.* Barry and Hartigan 1987, is concerned with the reconstruction of the evolutionary tree. We wish to contribute to the understanding of the time scale in given trees.

Thus consider some gene or protein, like the often studied α- or β-globins, and assume that there is, at each birth a probability p of mutation. In accordance with the infinite alleles interpretation each mutation is supposed to be unique. The underlying population is viewed as a supercritical, one-type branching process with the reproduction function μ, as in the preceding section. The expected number of first mutant progeny of an individual, *i.e.* of mutant individuals stemming from the individual and being the first mutants in their line of descent, is then

$$M(t) = p\sum_{n=1}^{\infty} q^{n-1}\mu^{*n}(t),\ q = 1-p,$$

within time t. Following Taib 1987, we clump an individual and all his non-mutant progeny into a *macroindividual*, born at the birth of the first individual of that allelic type, dead at the death of the last non-mutant descendant.

This creates a new branching process of macroindividuals, having the reproduction function M. The question how long a time ago the first in a sequel of n mutations occurred in the original population reduces to the question about Joe's n:th father in the population of macroindividuals, *cf.* also the more general analysis in Taib 1987, Chapter 7. To apply the answer of the latter to the former question it remains only to determine the Malthusian rate and the expectation and variance of the new process. But this is easy:

$$\hat{M}(\alpha) = p\sum_{n=1}^{\infty} q^{n-1}(\hat{\mu}(\alpha))^{n} = p\sum_{n=1}^{\infty} q^{n-1} = 1,$$

so that the Malthusian rate actually remains the same. But the other terms change drastically.

$$M_\alpha(dt) = e^{-\alpha t}M(dt) = p\sum_{n=1}^{\infty} q^{n-1}\mu_\alpha^{*n}(dt)$$

yields

$$\int_0^\infty tM_\alpha(dt) = p\sum_{n=1}^{\infty} q^{n-1}\int_0^\infty t\mu_\alpha^{*n}(dt) = p\sum_{n=1}^{\infty} q^{n-1}n\beta = \beta/p.$$

The classical variance decomposition applied to the sum of a geometrical number of independent random variables gives the variance

$$\upsilon^2/p + \beta^2 q/p^2.$$

Mutation probabilities are very small, so presumably this can be well approximated by just β^2/p^2, leading to the approximate time back to the first of n mutations as

$$Y + n\beta/p + \sqrt{n}\beta/pN(0,1).$$

Here Y, having the stable age distribution of the macroprocess, is the only problematic part. With a probability r, the probability of non-extinction of the original process, it is exponential (a), and with probability $1 - r$ it has the stable age distribution of a process whose life length has the distribution of the conditional time to extinction of the original process, given that it dies out. Presumably it can be disregarded in a situation with little p and large n. We thus arrive at the final approximation

$$n\beta/p + \sqrt{n}\beta/pN(0, 1).$$

In molecular genetics what is known is however not n but the number d of differences between two existing species. If the same mutation rate is assumed for both species, as is usually done, n in the above formula should be replaced by a random variable that is binomial with parameters d, 1/2. This and a central limit theorem for sums of a random number of random variables leads to our final formula:

The time back to separation between two species exhibiting at a certain amino acid chain d differences is, approximately,

$$d\beta/2p + \sqrt{d/2}\,\sqrt{\beta^2/2p + v^2/p + \beta^2 q/p^2}\,N(0, 1) \approx d\beta/2p + \sqrt{d/2}\,(\beta/2)\,N(0, 1)$$

Example. On a gene 80 differences have been found between two species. The time to their divergence has been estimated to 40 million years from other, *e.g.* fossile, evidence. If the generation time in the species considered is 10 years, equating $d\beta/2p$ with 40 million yields $p = 10^{-5}$. If this were known to be the true probability of mutation a 95% "postdiction" interval for the time to divergence would be $4 \times 10^7 \pm 1.3 \times 10^7$.

Alternatively one could argue that fossiles only yield lower bounds for the time to appearance of a species. Then we would conclude that

$$40 \times 10/p \;+ \sqrt{40} \times 10N(0, 1)/p > 4 \times 10^7,$$

and hence that

$$p < 1.36 \times 10^{-5}$$

with 99% confidence.

4. Concluding Remarks. This note sketches a branching process way of arguing about genetical evolution. It is very tentative. Serious application of this type of arguments would need more refined modelling. In particular the theory should incorporate multi-type individuals, so that mutation may possibly change the reproduction pattern. We hope to proceed to such work.

References

[1] Barry, D. and Hartigan J. A. (1987) Statistical analysis of hominoid molecular evolution. *Statist. Sci.* **2**, 191-210.

[2] Jagers, P. (1975) *Branching Processes with Biological Applications* . J. Wiley Sons, Chichester.

[3] Jagers, P. (1989) General branching processes as Markov fields. *Stoch. Proc. Appl.* 32.

[4] Jagers, P. and Nerman, O. (1984) The growth and composition of branching populations. *Adv. Appl. Prob.* 16, 221-259.

[5] Kimura, M. (1979) The neutral theory of molecular evolution. *Sci. American* 241, 94-104.

[6] Nerman, O. (1984) *The Growth and Composition of Supercritical Branching Populations on General Type Spaces.* Dep. Mathematics, Chalmers U. Tech. and Gothenburg U. 1984-18.

[7] Nerman, O. and Jagers, P. (1984) The stable doubly infinite pedigree process of supercritical branching populations. *Z. Wahrscheinlichkeitstheorie verw. Gebiete* 65, 445-460.

[8] Taib, Z. (1987) *Labelled Branching Processes with Applications to Neutral Evolution Theory.* Thesis, Department of Mathematics, Gothenburg U. To appear in Springer Lecture Notes in Biomathematics.

[9] Wilson A. C. et al. (1987) Molecular time scale for evolution. *Trends in Genetics* 3, 241-247.

POINT PROCESSES AND INFERENCE FOR RAINFALL FIELDS

MICHAEL J. PHELAN*
Princeton University

ABSTRACT

We treat the role of point processes in the statistical analysis of rainfall fields. We look in specific at Poisson-based models of the Le Camian type. These represent rainfall fields as a smoothing transformation of a Poisson random measure. We provide a number of examples of particular interest to hydrologists who analyze rainfall fields and we outline strategies for their assessment. Here we find a place for novel applications of statistical techniques, such as simulation, method of moments, nonlinear regression, and image and life history analysis.

1. **Introduction** Rainfall fields refer to the observed pattern of ground-level intensity of rain falling from precipitating cloud systems. Reflecting the nonequilibrium thermodynamic origins of such systems, these patterns manifest highly-structured distributions of rainfall over space. Our discussion here concerns modeling and inference for the temporal evolutions of these distributions.

The temporal evolutions of rainfall fields proceed from the changing thermodynamic conditions governing the phenomenon of precipitation. As part of a family of processes involved in global atmospheric circulation, precipitation is a process of complex geophysical origins. To model the dynamics of rainfall fields is obviously beyond the scope of present endeavors. In substitute, we treat phenomenological models of rainfall fields, that aim to describe the phenomenon as observed.

Nicolis and Prigogine (1977) note that the *observable* behavior of many nonequilibrium thermodynamic systems is well represented by a stochastic process. We take this view in modeling rainfall fields with a spatial stochastic process. In particular, we look at models that represent such fields as a smoothing transformation of a Poisson random measure. The general framework for these was introduced in Le Cam (1961). Nonetheless, it was not until after the GATE experiment of 1974 that Le Cam's work began to have impact. From this observational study of tropical rainfall over the Atlantic Ocean, meteorologists reported new understandings of the organization and structure of precipitating cloud

*Partially supported by Office of Naval Research Grant N00014-89-J-1830

systems; see for example Houze and Hobbs (1982). Since that time, Le Cam's work has influenced recent efforts to model rainfall fields including those in Smith and Karr (1985), Waymire, Gupta, and Rodriguez-Iturbe (1984), Rodriguez-Iturbe, Cox, and Eagleson (1986), Rodriguez-Iturbe, Cox, and Isham (1987, 1988), and Cox and Isham (1988).

We later recall the Le Camian representation of rainfall fields and discuss its principal characteristics. In providing a number of examples, we interpret the primitives in these representations in terms of the geophysical constructs of the raincell and the cloud cluster. These constructs have informed the discussions of rainfall events by meteorologist, and so provide the motivation behind the examples. As examples of particular interest to hydrologists who analyze rainfall fields, we include the WGR model in Waymire, et al. (1984) and the random-disks model in Cox and Isham (1988).

The complexity of these models and the phenomenon they represent compels the need for flexible strategies in their assessment. One such strategy, which we recall below, is outlined in Rodriguez-Iturbe, et al. (1988). This is a practical approach to parametric estimation that blends formal statistical procedures with informal physical judgements in promoting physically meaningful answers. To it, we add our own emphasis on the sample path analysis of rainfall fields as implemented in Phelan and Goodall (1989). This approach uses nonlinear regression and image analysis in data-analytic assessments of the ability of the model to represent observed rainfall fields. We emphasize that for any strategy to be successful it must stimulate the dialogue among observation, modeling, and inference.

This dialogue will often lead to fresh scientific inquiries as well as to novel applications of statistical techniques. To illustrate, we anchor our final discussion in the question of aging in rainfall fields. Here we introduce the aging function, which is a cumulative hazard function appearing as parameter in many models of rainfall. We address the problem of nonparametric estimation of such functions from observed rainfall fields, where we argue for the application of the techniques of life history analysis. In specific, we introduce a martingale-type estimator as a nonparametric estimator of the aging function in a class of point process models for rainfall fields. Finally, we integrate our approach to nonparametric estimation with the strategy for parametric estimation described above, and thus enhance its flexibility as an assessment regime.

2. Poisson Random Measures and Le Camian Representations The work in Le Cam (1961) develops a general framework for the stochastic description of precipitation using point processes and their transformations. Here we recall the main idea behind such descriptions and provide examples. Our intention is to draw attention to what we call Le Camian representations of rainfall fields.

Convective Raincells and Point Processes. In describing precipitating cloud systems, meteorologists refer to a geophysical construct known as the convective raincell. Briefly, the convective raincell denotes a localized region of high-intensity rainfall induced by convection or the exchange of heat from the earth to the atmosphere. In practice, we typically observe composites of such cells at a scale of about 4 km. At maturity, they may cover a surface area of 16 km^2 with rainfall at an intensity reaching 8 mm/hr. In the tropics for example, mature raincells are observed in groups called cloud clusters. They are usually connected by a layer of stratiform cloud. Aided by the winds, such precipitating systems horizontally traverse the lower atmosphere while depositing their water contents on the surface of the earth. We refer the reader to for example Houze and Hobbs (1982) for further details of the phenomenology of raincells in mesoscale descriptions of precipitative events.

In the importance of the convective raincell, Le Cam recognized a role for point processes in his stochastic descriptions of precipitation. The principal idea is that the occurrence of convective cells be identified with the atoms of a point process. In specific, let $(\Omega, \boldsymbol{H}, P)$ denote a probability space and $(E, \boldsymbol{E})$ a measurable space. Now let M and N denote random measures on $(E, \boldsymbol{E})$. We refer the reader to for example Karr (1986) for further mathematical details. We assume that N is a Cox process having directing measure M. Here in the context of modeling mesocale precipitation, N is the primitive that models the occurrence of convective raincells. For Le Cam, the specification of the space $(E, \boldsymbol{E})$ and the directing measure M is determined by the geophysical context and the design of the modeler.

We fix ideas and take E to be a product space of the form $T \times S \times \chi$. Here T denotes a subset of $\boldsymbol{R}$ and a time-index set, S denotes a subset of $\boldsymbol{R}^2$ and a planar region affixed to the surface of the earth, and χ denotes a vector space used in characterizing raincells by their water content, dispersion, velocity, etc. We assume that M is the fixed measure μ satisfying

$$\mu(dt, dx, dz) = c\,dt\,dx\,\pi(dz), \quad (t \in T,\ x \in S,\ z \in \chi), \tag{1}$$

where c is a positive constant and π denotes a probability measure on χ equipped with its measurable sets. Note that N is thus a Poisson random measure on $T \times S \times \chi$ with mean measure μ.

We say that the atoms of N identify convective raincells. That is, fix ω in Ω and suppose that (t, x, z) is an atom of N(ω) where $t \in T$, $x \in S$, and $z \in \chi$. Here (t, x) localizes the time and place of birth of a raincell. The vector z endows the raincell with a set of distinguishing characteristics such as its velocity and water content. According to equation (1), the model places raincells at random over time and space, whereas their characteristics vary independently over χ with common distribution π.

Remark. We mentioned earlier that convective raincells tend to occur in clusters. Le Cam addressed this issue in his proposal to construct the directing measure M to reflect the observed organizational structure of storm systems. For example, M may be chosen so that N is a cluster point process that clusters raincells in the manner observed in tropical storms. We refer the reader to Le Cam (1961) for the details of this part of the framework. In doing so, we nevertheless note that clustering among raincells is an important yet open issue in phenomenological modeling of rainfall. □

Rainfall Fields and Le Camian Representations Rainfall fields refer to the observed pattern of ground-level intensity of rain falling from a precipitating cloud system. Le Cam suggested that these be represented by a smoothing transformation of the point process N. In describing these representations, we proceed with our ideas fixed as above.

Our mathematical setting is the space $T \times S \times \chi$ equipped with its measurable sets, where we have defined the Poisson random measure N that generates raincells and their characteristics. We aim to model the temporal evolution of the intensity or rate of rain falling over S. For each $t \in T$ and $x \in S$, let $k(t, x)$ denote a positive, measurable function defined on $T \times S \times \chi$. Let $k = \{k(t, x), t \in T, x \in S\}$ denote the corresponding family of functions, where we assume k satisfies

$$\int_{T \times S \times \chi} \mu(du, dy, dz)\, k^2(t, x, u, y, z) < \infty, \tag{2}$$

for every $t \in T$, $x \in S$. Here μ is the mean measure belonging to N as specified at equation (1). Next, we introduce a spatial stochastic process $R = (R(t, x))$, $t \in T$, $x \in S$, where $R(t, x)$ satisfies

$$R(t,x) = \int_{T\times S\times\chi} N(du,\ dy,\ dz)\, k(t,x,u,y,z)\,, \tag{3}$$

Here we interpret $R(t,x)$ as the intensity of rain falling on position x at time t, so that the process R models the temporal evolution of the intensity of rain falling over S. Note that condition (2) guarantees the existence of at least the mean and variance of this process.

In the general framework of Le Cam, k is an example of a smoothing kernel, and R is thus an example of a smoothing transformation of N. A rainfall intensity process so obtained, we shall call a *Le Camian representation* of the rainfall field. Such representations appear in the examples below, where we add specific interpretations to the items introduced here.

Examples We illustrate the main ideas above and provide examples of particular interest to hydrologists who analyze rainfall fields. For simplicity in doing so, we omit the presence of a clustering mechanism, but we refer the reader to remark (1) below.

*a) The WGR Mode*l. This model is developed in Waymire, et al. (1984). Let T, S, and χ denote the sets $\mathbb{R}$ $\mathbb{R}^2$ and $\mathbb{R}_+$ respectively. We assume that N is a Poisson random measure on $T\times S\times\chi$ equipped with its Borel sets. For the purpose of interpretation, fix ω and suppose that $(t,\ xz)$ is an atom of $N(\omega)$. We then say that t and x is the time and place of birth of a raincell having water content or total intensity z. If the mean measure μ belonging to N satisfies equation (1), then π specifies the distribution of the raincells' water contents.

The WGR model includes a Le Camian representation of the rainfall intensity process. In particular, let a and b denote fixed, positive scalars and let v_0 denote a fixed vector in the plane. Here the smoothing kernel k satisfies: for each $t\in T$ and $x\in S$,

$$k(t,x,u,y,z) = z1(u\le t)\,ae^{-a(t-u)}\sqrt{2/\pi}\ be^{-bd(x,y+(1-u)v_0)}\,,$$

for every $u\in T$, $y\in S$, and $z\in\chi$. Here d(·,·) denotes the squared Euclidean distance between points in S. Under the smoothing kernel thus defined, the rainfall intensity process $R = R(t,\ x))$, $t\in T$, $x\in S$ is represented as in equation (3).

For each t and x, $R(t,\ x)$ denotes the total intensity of rain falling at $(t,\ x)$. It is the sum of the contributions of all the raincells born before time t. That is, fix ω and suppose that $(u,\ y,\ z)$ is an atom or raincell of $N(\omega)$. According to equation (3), $k(t,x,u,y,z)$ is the contribution of that raincell to the total intensity $R(\omega;\ t,$

x). In particular, by the action of k, its contribution is proportional to its total intensity z. To understand the proportion, we interpret $t - u$ as the age and $y + (t-u)\nu_0$ as the position of the raincell at time t. As it ages, the raincell thus propagates along the plane with velocity ν_0. We therefore see that, as a function of t, the raincell's contribution to $R(\omega; t, x)$ decays exponentially with its age and its Euclidean distance to x. Here we call a the rate of aging and b the coefficient of disperson. □

b) The Random-Disks Model. This model is developed in Cox and Isham (1988). As the name suggested, it envisions the coverage of a precipitating raincell to be a random disk in the plane. The initial placement of the disk is specified at birth, when the raincell is characterized by a radius, water content, duration, and planar velocity. A raincell becomes active at birth and it remains so for its duration. The ensemble of active raincells migrates s across the plane; the individual motions being at uniform rate along that raincell's velocity. In doing so, they deposit their water contents uniformly over their identifying disks and uniformly over their durations. Thereafter, the raincell vanishes.

We formalize the qualitative description above. Let T, S, and χ denote the sets R ,R^2 , $R^3_+ \times R^2$, respectively. We assume that N is a Poisson random measure on $T \times S \times \chi$ equipped with its Borel sets. For the purpose of interpretation, fix ω and suppose that $(t,x,(r,w,z,\nu))$ is an atom of $N(\omega)$, where $t \in T$, $x \in S$, $r, w, z \in R_+$, and $\nu \in R^2$. We then say that t and x is the time and place of birth of a raincell having radius or dispersion r, water content w, duration z, and velocity v. If the mean measure μ belonging to N satisfies equation (1), then t specified the joint distribution of the raincell' redii, water contents, durations, and velocities.

For each time t and site x, let $R(t,x)$ denote the total intensity of rain falling at (t,x). According to the model, this is given by the sum of the contributions of those active raincells that cover the site x. Here we give its Le Camian representations. Let k denote the smoothing kernel satisfying: for each $t \in T$ and $x \in S$,

$$k(t,x,u,y,r,w,z,\nu) = w1(u \leq t)\, 1\,(d(x,y + (t - u)\nu) \leq r)1(z \geq t - u),$$

for every $u \in T$, $y \in S$, $r,w,z \in R_+$, and $\nu \in R^2$. Now we define $R(t,x)$ according to equation (3), and thus obtain the desired representation of the rainfall field $R = (R(t,x)$, $t \in T$, $x \in S$. In leaving the specifics of the interpretation of the action of k to the reader, we ask him to verify that the representation of R given here is (stochastically) equivalent to that of equaiton (8) in Cox and Isham (1988). □

Remarks

(1) *Clustering*. The models described above admit generalizations involving higher-order clustering. This entails constructing a directing measure M that subordinates the point process N. For example, M itself may be a smoothing transformation of an initial point process. The action of the initial process is to generate cluster potentials, so that M may direct N to cluster raincells in the vicinity of these. Such a mechanism may well model the appearance of the cloud clusters mentioned earlier. For example, this device is used in Cox and Isham (1988) to induce either Bartlett-Lewis or Neyman-Scott type cluster processes.

(2) *Geometry and Kinematics*. The choice of smoothing kernel reflects or imposes a choice of geometry and kinematics for the underlying storm system. In the geometry of the WGR model, for example, each raincell is identified with a continuum of Gaussian surfaces. This has obvious ramifications for the spatial distribution of rain falling from that cell. In contrast, the geometry of the random-disks model distributes rain over a migrating coverage process, so that each raincell deposits water uniformly over a random set in the plane. Nevertheless, the two models maintain essentially the same kinematics. The raincells move at uniform rate along a specified velocity. The difference lies in that they do so in lockstep fashion in WGR, whereas random-disks allows for individual variations in these motions. □

3. Strategies in Assessment Stochastic descriptions of precipitation are useful from a number of viewpoints. On the one hand, they offer a ready framework to guide the hand of the data analyst. They suggest ways of examining observations and discovering relationships. On the other hand, they can be an integral part of scientific inference about precipitation or a workable tool in the engineering of water resources. Hydrologists, for example, use stochastic models in forecasting and simulating precipitation as part of the hydrological cycle. They may require only that such descriptions preserve averages or provide realistic patterns in rainfall activity. Nevertheless, in assessing stochastic descriptions of precipitation, we do well to remember the physical origins of the phenomenon in designing strategies and in judging the descriptive capabilities of the model.

In this section, we describe two broad strategies for the assessment of point process models of rainfall fields. The first is basically a parametric approach as described in Rodriguez-Iturbe, Cox, and Isham (1988). The method is well suited for assessing a model's ability to preserve averages in rainfall activity. The second strategy is a data-analytic approach to the sample path analysis of rainfall fields as implemented in Phelan and Goodall (1989). As such, it is well suited for assessing a model's ability to provide realistic patterns of rainfall. The two approaches are complementary, and together they offer a practical strategy for assessing Le Camian representations of rainfall fields.

I. The Method of Primary Features. This method is both a method of fitting parameters in a rainfall model and of assessing the fitted model. Respecting the earnest conceptual and computational problems entailed by the likelihood approach to parametric estimation from rainfall fields, the procedure recommended here involves solving a system of simultaneous equations drawn from a set of primary features of the data. Typically, only a subset of such features are needed to fit a finite dimensional parameter. Once fitted, therefore, the model is assessed against the subset of primary features that remain.

We fix ideas and consider the random-disks model as described in the previous section. There the mean measure μ depends on two parameters; the rate c and the distribution π. Recall that π denotes the distribution of the raincells' characteristics including their radii, water contents, durations, and velocities. For the present illustration, we assume that π satisfies

$$\pi(dr, dw, dz, dv) = ae^{-ar}\, be^{-bw}\, de^{-dz}\, drdwdz\, Q(dv), \tag{4}$$

for every $r,w,z \in \mathbf{R}_+$, and $v \in \mathbf{R}^2$. Here a, b, and d denote fixed parameters and Q satisfies

$$Q(dv) = (\beta(\beta\| \, v \, \|)^{\alpha-1}\, e^{-\beta\| \, v \, \|} / \, \Gamma(\alpha))\delta_0,$$

where $\| \, v \, \|$ denotes the length of v, α and β are fixed, positive parameters, and δ_0 is the direct measure concentrated at the fixed vector v_0 of unit length in the plane. This choice of π entails the independence of the raincells' characteristics. Moreover, the radii, water contents, and durations are exponentially distributed with parameters a, b, and d, respectively. Finally, the velocities are concentrated along the direction v_0, while their magnitudes are Gamma distributed with shape parameter α and scale parameter β. We have therefore specified a model of 7 parameters, namely a,b,c,d,α,β, and v_0, which we hereafter denote by the symbol θ.

The problem is to estimate θ from a record of observed rainfall activity. The method of primary features resolves the problem as follows. Typically, the data will consist of partial observations from the rainfall intensity process R, perhaps drawn from a network of rain gauges placed in a catchment. Let **G** denote a set of primary features of the data. We require that the corresponding model features be available either theoretically or by simulation. As examples of features in **G**, we take means, variances, and spatial-temporal autocorrelations of rainfall intensity, as well as derived distributions of say dry periods, level crossings, or extreme values in rainfall activity. Next, let **F** denote a subset of **G** to be used in fitting θ. We therefore require that **F** contain at least as many features as the dimension of our parameter. Now for each $f \in F$, let $D(f)$ denote the value of f as determined by the data. Similarly, let $M(\theta;f)$ denote the value of f as determined by the model. For example, suppose f denotes the proportion of time it is dry at a fixed site. Then $D(f)$ is the observed fraction of time it is dry there, and $M(\theta;f)$ is $\exp(-2c\pi/da^2)$ as determined by equaiton (2) in Cox and Isham (1988). The method of primary features is to solve for θ by solving the system of simultaneous equations satisfying

$$D(f) = M(\theta;f)\,,\; f \in F\,, \tag{5}$$

provided there is a unique solution.

The choice of features to fill **F** will be guided by many criteria. The foremost is the observational scheme, as we require data-based values for every feature in **F**. Otherwise, we look for features that will provide a sensitive and balanced test of the model's capabilities. For example, we select features that are relatively uncorrelated and have small sampling variability. Also, we select features that are available at a range of scales, this reflecting the physical importance of scaling principles in rainfall fields. Overall, we aim to achieve a nice blend of formal statistical thinking with informal physical judgment. For example, we may incorporate auxiliary information and appeal to physical measurements of the ambient winds to determine ν_0. These measurements may then be used to help determine the equations above, say, by fixing an initial value to the mean α/β in the distribution of raincells' velocities.

In the method of primary features, the fitted model is assessed in two ways. The first is cross-validation, where the model-based values of the features in **F** are compared to data-based values as computed from a hold-out sample. Alternatively or in addition, if data-based values of the features in **F** are available over a range of scales of aggregation, say hourly, daily, and monthly amounts of rainfall, then model-based values are assessed across this range. In either case, we assess the capability of the model to reproduce the set of primary features used in fitting θ. On the other hand, the second way in which the model is assessed is to consider those primary features in $\mathbf{F}^c$, where $\mathbf{F}^c$ denotes the complement of **F** in **G**. Here we assess the capability of the model to extrapolate to the set of primary feature not used in fitting θ. For example, suppose the distribution of extreme values at a fixed site is derived (at least approximately) from the model and placed among the features in $\mathbf{F}^c$. Then a formal or diagnostic test of the model is made by comparing observed extremes to the extremes expected by the model evaluated at the value of the fitted parameter.

We believe that the method of primary features is a workable and flexible strategy for the assessment of point process models of rainfall fields. We find that it is particularly successful at selecting models that well preserve averages in rainfall activity. This is due in part to the computational ease of using various statistical moments as primary features. Nevertheless, it is of certain value to the community of hydrologists and for illustrations of the methodology, we refer the reader specifically to Rodriguez-Iturbe, Cox and Isham (1987, 1988).

II. A Method of Sample Path Analysis. This method aims to probe the representational capabilities of stochastic descriptions of precipitation, such as the Le Camian representations of rainfall fields. The method proceeds through a sample path analysis of an observed rainfall field in the context of a particular representation. The basic exploratory principal is that the observed data may be viewed as the composition of a fitted representation with a residual. Here the fitted representations are probed for their ability to provide realistic patterns of rainfall and the residuals are probed for systematic departures from the patterns desired. The procedures involved range from nonlinear regression and image analysis to time-series analysis and cross-validation against independent analyses of similar or related rainfall events.

We fix ideas and consider the WGR model as described in the previous section. We do, however, introduce a modest generalization. First, let χ denote the set $R_+^3 \times \mathbb{R}^2$. We assume that N is a cluster point process on $T \times S \times \chi$, although the exact clustering regime is not central to this treatment. Here the raincells are endowed with four random characteristics; namely water content, aging rate, dispersion, and velocity. Thus the water contents are random as before, but the parameters a, b, and v_0 appearing in the WGR smoothing kernel are additionally taken to be random variables. This leaves the functional form of the smoothing kernel itself as the parameter determining the Le Camian representation in the model.

The problem is to assess the choice of smoothing kernel from a record of observed rainfall fields. Typically, the data consist of a time series of radar images drawn from the partial observation of the rainfall intensity process R. As an example of such data, we take the GATE rainfall fields as converted by Hudlow and Patterson (1979). These fields refer to a time series of radar images depicting derived rates of rain falling over a disk-shaped region of the Atlantic Ocean during the summer of 1974. The issue is thus the extent to which the rainfall intensity derived from the WGR model produces sample paths that represent the spatial evolution exhibited in this time series.

To fix ideas further, let $Y = (Y(t,x)), t \in I$ and $x \in D$ denote the time series of observed rainfall rates. Here I and D denote respectively the summer period of 1974 and the observational disk over the Atlantic Ocean. We view the data Y as a composition satisfying

$$Y(t, x) = R(t, x) + Z(t, x),$$

for every $t \in I$ and $x \in D$. Here the $R(t, x)$ denote the WGR representations of $Y(t,x)$ (see equation (3)) and the $Z(t,x)$ denote the corresponding residuals. The principal strategy of the present method of assessment is to probe the sample paths of the triplet of processes Y, R, and Z for evidence of the descriptive capabilities of the model.

The method of sample path analysis proceeds as follows:

(1) *The Fit.* The first step is to obtain a fitted version of the model-based representation. Recall that R is a function of the point process N and the smoothing kernel k. Since our focus is set upon assessing the choice of k and N cannot be observed directly, we look for a suitable state estimator $\hat{N}$, say, of N. A general theory for handling such problems is described in section 3.3 of Karr (1986). As a practical approach, we recommend finding a least-squares fit of N by minimizing the square-error loss between R and Y over the range of observation. In our case, this involves estimating the placements and the characteristics of those raincells generating Y. This may be handled using nonlinear regression. That is, we introduce a vector-valued "parameter" consisting of those atoms of N whose placements lie in $I \times D$; namely the placements and the characteristics of all the existing raincells perceived to be generating rainfall over the window of observation. We then solve for the minimizing value of that vector using the Gauss-Newton procedure described in Bates and Watts (1988).

(2) *Image Analysis.* Let $\hat{N}$ denote a suitable fit of the raincells and their characteristics. On substituting $\hat{N}$ for N in equation (3), we obtain a fitted ver-

sion $\hat{R}$, say, of the model-based representation of Y. Of course, the patterns of rainfall manifest by $\hat{R}$ are determined by the choice of the WGR kernel, and thus these patterns can be explored for the representational capabilities entailed in this choice. Generally speaking, we look for agreement in shape, organization, magnitude, and temporal evolution between the fields in the fitted image and those in the images of data. This may be handled informally by comparing animations of the two time series or by detailed data-analysis of individual images. Of course, we recommend doing both.

Next, we look at the least-squares residuals $\hat{Z}$ obtained by subtracting $\hat{R}$ from Y. This yields a time series of images depicting departures in the data from the fitted WGR representation. Here we examine the temporal evolution of the distribution of residuals with particular regard to their spatial homogeneity. That is, we look for reasonably Gaussian (i.e. structureless) distributions without apparent anomalies; this may be handled using normal probability plots. Secondly and perhaps more importantly, we examine the images of residuals for their adherence to spatial homogeneity using animation or detailed analysis of individual images. For example, the presence of inhomogeneities may reflect an inability of the model to properly fit peak intensities located at the raincells' centers or an inability to reproduce the stratiform precipitation supported by many cloud systems. By such an examination, we ascertain when in the life cycle of a storm system the WGR model gives an adequate description of precipitation and, moreover, when it fails to do so.

(3) *Decoupling and Time Series Analysis.* As a further exploration of the data and the model, we recommend that the analysis above be decoupled in time. That is, we are analyzing a time series of spatial distributions of rainfall over D, where, for each $t \in I$, $Y(t) = (Y(t,x))_{x \in D}$ denotes the observed distribution at time t. Here we recommend analyzing $Y(t)$, separately for every t, using the techniques above. In this sense, we decouple the analysis in time.

The aim of the decoupled analysis is for example to partially separate the choice of geometry from the choice of kinematics entailed in the choice of the smoothing kernel (see remark (2) of the previous section). Thus, for each t in the decoupled analysis, we decouple the WGR representation as follows. Let N_t denote a point process on $S \times R_+^2$. Here, if (x_t, z_t, b_t) denotes an atom of N_t for some ω, then x_t denotes the raincell's position and z_t and b_t denote its intensity and coefficient of dispersion. Now we introduce the decoupled smoothing kerned k' satisfying

$$k'(y;x_t,z_t,b_t) = z_t \sqrt{2/\pi}\; b_t \; e^{-b_t d(y,x_t)},$$

for every $y \in S$. Note that the decoupled kernel k' preserves only the spatial component or geometry in the starting WGR kernel. We have, for example, absorbed the exponential aging term into the time-dependent intensity z_t.

The decoupled analysis proceeds by obtaining a fit $\hat{N}_t$, say, to N_t, a fitted representation $\hat{R}_t$ of $Y(t)$, and the corresponding residual $\hat{Z}_t$. Here $\hat{R}_t$ and $\hat{Z}_t$ are put to the individual image analyses mentioned above, again emphasizing representational capability and spatial homogeneity. In addition, the $\hat{N}_t$ provide a times series of fitted raincells and their characteristics. These may be analyzed as a multivariate time series for consistency with the hypothesized model and historical analyses of similar rainfall events. In the case of the former, one can examine the choice of exponential aging as well as the chosen kinematic mechanism describing the motions of the raincells.

In closing, we remark that the analyses above depend to some extent upon subjective evaluations as well as a variety of informal physical judgements. For example, detecting raincells in radar images is to some degree a subjective operation, but this can be achieved with the help of sound data-analytic definitions of raincells as drawn from the meteorological literature. In fact, without the use of such informal physical judgements one may not be able to fix suitable starting values in the nonlinear regressions cited above. Nevertheless, some of these judgements may ultimately find support in the stability of the resulting least squares fits. Finally, we believe that the method of sample path analysis is a strategy that is complementary to the method of primary features for assessing Le Camian representations of rainfall fields. For an illustration of the methodology, we refer the reader to the analysis in Phelan and Goodall (1989).

4. *Aging Functions and Their Nonparametric Estimation.* The purpose of this section is to discuss the role of aging in rainfall fields. Our principal aim is to introduce the aging function, which is a cumulative hazard function appearing as parameter in many models of rainfall. We define a nonparametric estimator of the aging function and sketch its properties. In closing, we suggest integrating the estimator into the method of primary features as described in the previous section.

Aging in Rainfall Fields. The process of aging in convective raincells appears to be integral to the evolution of rainfall fields. For example, aging has been found empirically to effect the propagation of storms, the structure of turbulence in rainfall fields, as well as the shape of the spatial distribution of rainfall intensity.

For example, Houze (1981) discusses the importance of the aging of older raincells in the creation of new raincells at the leading edge of a storm system. He describes the resulting interplay between downdrafts and updrafts that appear to serve the propagation of the system. On the other hand, Zawadski (1973) drew a connection between the decay of the empirical autocovariance function of an observed rainfall field and Taylor's hypothesis of turbulence in such fields. Briefly, he showed that the expected duration of a convective raincell appears to define a cutoff for the approximate validity of this hypothesis for rainfall fields. Finally, using the method of sample path analysis outlined above, Phelan and Goodall (1989) assessed the WGR model against GATE rainfall fields. We found that the model fit the data reasonably well during the earlier stages of a storm, but that as the convective raincells that were supporting the storm neared the end of their effective lives, the model lost its ability to describe the patterns of observed rainfall. It thus appears that the geometry and kinematics of an aging storm is notably different than that of a developing or mature storm.

The importance of aging in these and other treatments of observed rainfall fields suggests the need to better understand the process of aging in convective raincells. One way to do this is in the context of a statistical investigation of the durations of such cells. The remainder of our discussion illustrates this approach by treating the problem of survival analysis of convective raincells from partially observed rainfall fields.

Aging Functions in Rainfall Models. An aging function is simply a cumulative hazard function belonging to a positive random variable. Nevertheless, we prefer the simpler term aging function. One reason for this preference lies in the role such functions play in the story of aging in rainfall fields; a fuller treatment of this role is to be found in Phelan (1989). We define an aging function.

Definition *Aging function.* The function $A = (A(t))$, $t \geq 0$ is said to be an aging function if A is an increasing, continuous function satisfying $A(0) = 0$ and $A(t)$ tends to infinity with t. □

Note that we have chosen to restrict our treatment to continuous aging functions for the mathematical convenience of this choice below.

We use aging functions here to parametrize the distribution of positive random variables. In specific, let X denote a positive random variable of distribution ψ. The random variable X is said to have aging function $A = A(t))$, $t \geq 0$ if

$$1 - \psi(t) = e^{-A(t)} \text{ or } \psi(dt) = A(dt)e^{-A(t)}, \quad t \geq 0,$$

The aging function is thus minus the natural logarithm of the survival curves belonging to X, which, in the context of survival analysis, is also called the cumulative hazard function.

We give an example. Let X denote a positive random variable having Pareto distribution of shape parameter b and scale parameter c. The aging function belonging to X then satisfies

$$A(t) = b\ln(1 + ct/b), \quad t \geq 0.$$

This aging function entails that X has a long-tailed distribution relative to, say, the distribution of an exponential random variable. Nevertheless, as b tends to infinity, $A(t)$ tends to ct for every t, where the latter is the aging function belonging to an exponential random variable of parameter c.

Aging functions appear as statistical parameters in many stochastic models of rainfall fields. As an illustration, we fix ideas and introduce the aging function to the random-disks model as described in sections 3 and 4. Recall that N is a Poisson random measure on $T \times S \times \chi$, where χ is the product space $R_+^3 \times R^2$. Here we simplify the model in assuming that the mean measure μ belonging to N satisfies

$$\mu(dt, dx, dr, dw, dz, dv) = dt dx dr e^{-r} dw e^{-w} dz e^{-z} \delta_0,$$

for every $t \in T, x \in S, r, w, z \in R_+$, and $v \in R^2$, where δ_0 is the direct measure concentrated at the fixed velocity v_0. The raincells thus move lockstep along the fixed velocity v_0. Moreover, their radii, water contents, and durations are independent exponential random variables of unit parameter.

Next, let $A = (A(t))$, $t \geq 0$ denote an aging function and let f denote the measurable mapping satisfying

$$f(t, x, r, w, z, v) = (t, x, r, w, A^{-}(z), v),$$

for every $t \in T, x \in S, r, w, z \in R_+$, and $v \in R^2$. Here A^{-} denotes the functional inverse of A. Now consider the point process $\tilde{N}$ given by $N o f^{-1}$, where f^{-1} denotes the inverse image of f. According to definition 1.36 in Karr (1986), $\tilde{N}$ is a measurable mapping of N by f. Moreover, $\tilde{N}$ is a Poisson random measure on

$T \times S \times \chi$. Its mean measure $\tilde{\mu}$ is $\mu o f^{-1}$ and satisfies

$$\tilde{\mu}(dt,dx,dr,dw,dz,dv) = dtdxdre^{-r}dwe^{-w}A(dz)e^{-A(z)}\delta_0,$$

for every $t \in T, x \in S, r, w, z \in R_+$, and $v \in R^2$. This follows easily from the nature of the transformation f. In specific, fix ω and suppose (t, x, r, w, z, v) is an atom of $N(\omega)$. Then, by definition, the point $(t, x, r, w, A^{-}(z), v)$ is an atom of $\tilde{N}(\omega)$. Now, it is well known that the mapping $z \to A^{-}(z)$ transforms the exponentially distributed durations among the atoms of N to durations among the atoms of $\tilde{N}$ having aging function A. This, or course, implies the desired result.

According to the argument above, $\tilde{N}$ generates raincells whose durations have aging function A. Obviously, we can enrich the mapping f by introducing transformations on the remaining characteristics of the raincells, but this would distract from our present purpose. That is, we view A as an infinite dimensional parameter in the model above and address its nonparametric estimation.

Nonparametric Estimation. We consider the nonparametric estimation of the aging function A appearing in the random-disks model described above. In this consideration, we describe a suitable observational scheme that allows for the partial observation of a set of convective raincells and their activities. We then define a martingale-type estimator of A that is to be drawn from these partial observations.

We imagine that we are positioned at a fixed weather radar station, such as the one presently operating off Darwin, Australia. Over a period of time, storm systems pass in the vicinity of our instruments, which record the activities of the convective raincells. We specifically assume that our instruments track the trajectories of the raincells as they pass through the field of view. An objective method for doing so is described in Rosenfield (1987).

More formally, let $T_0 \subset T$ denote an interval of time and let $S_0 \subset S$ denote a disk of fixed radius in the plane. Here T_0 denotes the period of observation and S_0 denotes the field of view of our instruments, so that $T_0 \times S_0$ denotes the coverage window of our observational scheme. Our scheme is to observe the trajectory of a raincell in S_0, for every raincell that enters S_0 during the period T_0.

We assume that raincells and their characteristics are being generated by the

Poisson random measure $\tilde{N}$ on $T \times S \times \chi$, where $\tilde{N}$ is described above. We identify $\tilde{N}$ with the countable sequence $(T_n, X_n, R_n, W_n, Z_n, v_0)$, $n \geq 0$ of $T \times S \times \chi$ - valued random variables. For the raincell having label n, (T_n, X_n) denotes its time and place of birth in $T \times S$, R_n, W_n, and Z_n respectively denote its radius, water content, and duration, and v_0 denotes its fixed velocity. According to the kinematics of the random-disks model, the trajectory of the raincell traces a line segment L_n, say, in S. According to our observational scheme, we observe that part of L_n contained in S_0. Figure 1 depicts the four contingencies in the partial observation of L_n, depending on the position of the raincell at birth and at death.

More formally, for each n, let L_n denote the line segment in S traced by the trajectory of the raincell labeled n Then L_n satisfies

$$L_n = \{x \colon x \in S, x = X_n + tv_0 \text{ or some } t \in [0, Z_n] \}.$$

Let $l(L_n)$ denote the length of L_n. Clearly, we have

$$Z_n = l(L_n)/\|v_0\|,$$

for every n, provided $\|v_0\| > 0$. Thus, by observing the trajectory of the raincell, we observe its duration. The observational scheme, however, allows only the partial observation of the L_n, and thus it allows only the partial observation of the Z_n. That is, we observe only the subtrajectory $S_0 \cap L_n$, see figure 1, and therefore we observe only the censored duration $\tilde{Z}_n$ satisfying

$$\tilde{Z}_n = l(S_0 \cap L_n)/\|v_0\|.$$

According to the model, the Z_n have aging function $A = (A(t))$, $t \geq 0$. Here we propose an estimator of A to be drawn from the $\tilde{Z}_n$; the censored durations. For each n, let δ_n denote the indicator $1(Z_n = \tilde{Z}_n)$; this indicator is one for any raincell whose trajectory is fully contained in S_0. Let M denote the random variable satisfying

$$M = \sum_{n \geq 0} 1([T_n, T_n + Z_n \cap T_0 \neq \phi, S_0 \cap L_n \neq \emptyset).$$

Here M is the random number of raincells whose trajectories enter our field of view. If for example T_0 is a bounded interval, then M is finite almost surely.

Now let (N_t), $t \geq 0$ and (R_t), $t \geq 0$ denote the stochastic processes satisfying

$$N_t = \sum_{n=1}^{M} 1(\tilde{Z}_n \leq t,\ \delta_n = 1) \text{ and } R_t = \sum_{n=1}^{M} 1(\tilde{Z}_n \geq t).$$

These processes are respectively the analogue of the survival counting process and the risk process referred to in the analysis of censored survival data; see for example Andersen and Borgan (1985). Finally, we define the stochastic process $\tilde{A} = (\tilde{A}_t)$, $t \geq 0$ satisfying

$$\tilde{A} = \int_0^t \frac{dN_s}{R_s}.$$

Here we propose $\tilde{A}$ as a nonparametric estimator of the aging function A.

Properties of the Estimator $\tilde{A}$. The estimator $\tilde{A}$ is in essence a Nelson-Aalen estimator of the aging function A. It may be justified as a nonparametric maximum likelihood estimator (NMLE) or as a martingale estimator. In the case of the former, one writes down the appropriate likelihood for the censored durations as a function of the infinite dimensional parameter A. Then one appeals to a generalized principle of maximum likelihood to derive $\tilde{A}$ as the NMLE in the manner of argument given for example in Johansen (1978).

On the other hand, $\tilde{A}$ is also a martingale estimator. That is, let $G = \left(\underset{=}{G}_t \right)$, $t \geq 0$ denote the internal history generated by the counting process (N_t), $t \geq 0$. Then the process formed of the difference between $\tilde{A}$ and A, namely $\tilde{A} - A = (\tilde{A} - A(t))$, $t \geq 0$, is a semimartingale relative to G. Although such estimators have been studied extensively in for example Andersen and Borgan (1985), their application to the analysis of rainfall fields is a novel one.

Whether one views $\tilde{A}$ as an NMLE or a martingale estimator, the martingale property is useful in developing its statistical properties. For example, we consider the asymptotic distribution of $\tilde{A}$ as the observational period T_0 grows unboundedly. In this consideration, we expect that $\tilde{A}$ - A (suitably normalized) converges weakly to a Gaussian process. Since $\tilde{A}$ is a martingale estimator, this expectation can be proved by appeal to the appropriate martingale central limit theorem.

A complete treatment of the statistical properties of $\tilde{A}$ are to appear in a Phelan (1990). There we treat a broader range of observational schemes and stochastic models of greater generality than the one described above.

The estimator $\tilde{A}$ and The Method of Primary Features. In order to integrate the estimator $\tilde{A}$ into the method of primary features, we recommend that the aging function A be chosen as one of the primary features. That is, we identify A with the model-based value of the "aging-function feature." Typically A or $A(\theta)$ will depend on one or two parameters in the model, such as the shape and scale parameter determining the Pareto aging function exhibited above. Next, we identify the data-based value of the aging-function feature with the nonparametric estimator $\tilde{A}$. Finally, we incorporate the pair $(A(\theta),\tilde{A})$ directly into either the estimation or the post-estimation assessment procedure developed in the method of primary features. On the one hand, $\tilde{A}$ may be used to determine the parameter θ in the simultaneous equation shown at (5). Otherwise, suppose an estimate $\hat{\theta}$ of θ is obtained from (5) without the help of $\tilde{A}$. We then use $\hat{\theta}$ to form the model-based parametric estimator $A(\hat{\theta})$ of $A(\theta)$ and we compare $A(\hat{\theta})$ against $\tilde{A}$ as a partial, nonparametric assessment of the fit of the model.

5. Concluding Remarks. We argued for the use of spatial stochastic processes as phenomenological models of rainfall fields. In this spirit, we treated the stochastic descriptions in Le Cam (1961). We described Le Camian representations of rainfall fields and cited two examples of interest to hydrologists who analyze such fields.

We referred to the need of flexible strategies in the assessment of Le Camian representations of rainfall. In doing so, we outlined two strategies of assessment; one called the method of primary features and the other a method of sample path analysis. We argued that the two offer complementary approaches to the problem of assessment. We remark here that each strategy is made possible by the increase in sophistication of both data and data-handling capability available today.

We pointed out that the dialogue among observation, modeling, and inference leads to fresh scientific inquiries as well as to novel applications of statistical technique. In illustration, we introduced a nonparametric estimator of an aging function in a rainfall model by way of a martingale estimator typically found in the literature on life history analysis. We argued for the integration of this estimator into the method of primary features.

We remark, finally, that the Le Camian framework for the stochastic description of precipitation remains to be fully explored. For example, little is understood of how well these models satisfy the evolution equations that describe the physical dynamics of precipitation. Notwithstanding, more work is needed in the theory and application of such descriptions in the statistical analysis of observed rainfall fields.

References

[1] Andersen, P. K. and Borgan, O. (1985). "Counting process models for life history data: a review." *Scand. J. Statist.* **12**, 97-158.

[2] Bates, D. M. and Watts, D. G. (1988). *Nonlinear Regression Analysis and Its Applications*, John Wiley & Sons, Inc., New York.

[3] Cox, D. R. and Isham, V. S. (1988). "A simple spatial-temporal model of rainfall," *Proc. R. Soc. Lond.*, **A 915**, 317-328.

[4] Houze, R. A. (1981). "Structure of atmospheric precipitation systems: a global survey," *Radio Science*, **16**, 671-689.

[5] Houze, R. A. and Hobbs, P. V. (1982). "Organization and structure of precipitating cloud systems," *Adv. Geophys.*, **24**, 225-315.

[6] Hudlow, M. D. and Patterson, P. (1979). GATE Radar Rainfall Atlas, NOAA Special Report.

[7] Johansen, S. (1978). "The product limit estimator as maximum likelihood estimator." *Scand. J. Statist.* **5**, 195-199.

[8] Karr, A. (1986). *Point Processes and Their Statistical Inference*, Marcel Dekker, Inc., New York.

[9] Le Cam, L. M. (1961). "A stochastic description of precipitation," in *Proceeding of the Fourth Berkeley Symposium on Mathematical Statistics and Probability* (J. Neyman, ed.), 3, 165-186, Berkely, CA.

[10] Nicolis, G. and Prigogine, I. (1977). *Self-Organization in Nonequilibrium Systems*, John Wiley and Sons, New York.

[11] Phelan, M. J. (1991). "Aging functions and their nonparametric estimation in point process models of rainfall fields," to appear in *Statistics in Environmental and Earth Sciences*, P. Guttorp and A. T. Walden, eds., Edward Arnold Limited, Kent, England.

[12] Phelan, M. J. and Goodall, C. R. (1989). "An assessment of a generalized WGR model for GATE rainfall. *J. Geophys. Res., Atmos.*

[13] Rodriguez-Iturbe, I., Cox, D. R. and Eagleson, P. S. (1986). "Spatial modelling of total storm rainfall, "*Proc. R. Soc. Lond. A*, **403** 27-50.

[14] Rodriguez-Iturbe, I., Cox, D. R., and Isham, V. (1987). "Some models for rainfall based on stochastic point processes, "*Proc. R. Soc. Lond. A*, **410**, 269-288.

[15] Rodriguez-Iturbe, I., cox, D. r. and Isham, V. (1988). "A point process model for rainfall: further developments, "*Proc. R. Soc. Lond. A.*, **417**, 283-298.

[16] Rosenfeld, (1987). "Objective method for tracking convective cells as seen by radar," *J. Atmos. and Oceanic Tech.*, **4**, 422.

[17] Smith, J. A. and Karr, A. F. (1985). "Parameter estimation for a model of space-time rainfall," *Water Resource Res.*, **21**, 1251-1257.

[18] Waymire, E., Gupta, V. K., and Rodriguez-Iturbe, I. (1984). "Spectral theory of rainfall intensity at the meso-β scale," *Water Resource Res.*, **20**, 1453-1465.

[19] Zawadzki, I. I. (1973). "Statistical properties of precipitation patterns," *J. Appl. Meteorol.*, **121**, 459-472.

Figure 1. Four contingencies in the partial observation of L_n. Here S_0 denotes a disk in the plane, v_0 points in the horizontal direction, x denotes the position of the raincell at birth, and 0 denotes its position at death.

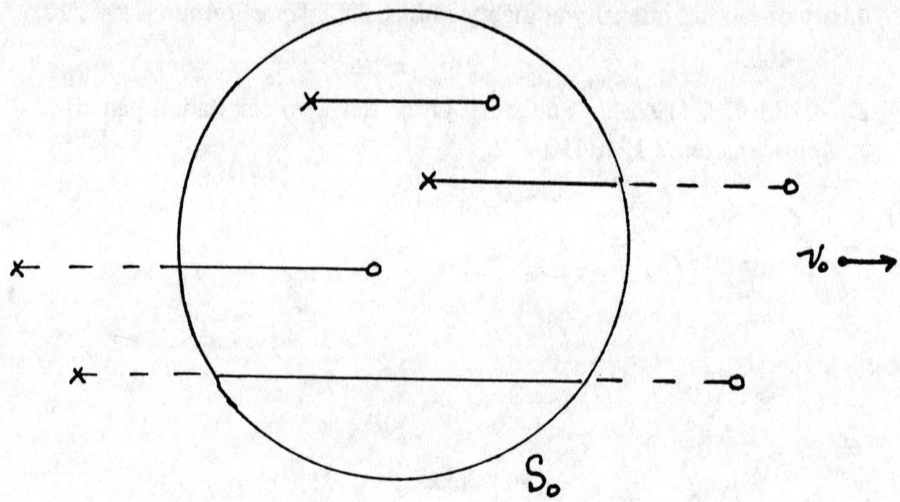

A CONSTRUCTION FOR PROCESSES WITH HISTORY-DEPENDENT TRANSITION INTENSITIES

P. Whittle
University of Cambridge

Summary

At the Sheffield meeting in honour of Joe Gani I presented the material of Whittle (1990); this was essentially Theorem 1 below and applications of it in network contexts. In the ensuing discussion John Bather raised an interesting point, which I now pursue.

1. Introduction. Suppose that a continuous-time Markov process has state variable x and infinitesmal generator T. Its equilibrium density $\pi(x)$ relative to a suitable measure μ then satisfies $T'\pi = 0$, where T' is the operator adjoint to T in that $\int(\pi Tf)\mu(dx) = \int(fT'\pi)\mu(dx)$ for functions $\pi(x)$, $f(x)$. Consider now a family of such processes parametrised by a vector $\nu = \{\nu_j\}$ for which the generator has the form

$$T(\nu) = \sum_n \nu_j T_j \tag{1}$$

Here the T_j are a set of fixed generators and ν_j a set of variable parameters, scalar and non-negative. One can regard T_j as the generator of transitions by a particular mode (the jth mode); these different modes being weighted differently as we vary the parameters ν_j. We shall refer to the vector $\nu = \{\nu_j\}$ as the rate vector; its essential property is that it enters linearly into the generator.

Let the family of processes generated as ν varies in some given set N be denoted F. Let us also suppose that the equilibrium density $\pi(x|\nu)$ of the process with intensity (1) is unique and known, for all ν in N. Note that prescription (1) includes also the non-homgeneous case

$$T(\nu) = T_0 + \sum_{j\neq 0} \nu_j T_j \tag{2}$$

This corresponds to the case when $\nu_0 \equiv 1$ for ν in N.

Consider now the mixed density

$$\tilde{\pi}(x) = \int \pi(x|\nu)\phi(d\nu) \tag{3}$$

where ϕ is a probability measure on N. Consider the process with generator

$$\tilde{T} = \sum_j \tilde{v}_j(x) T_j \tag{4}$$

where

$$\tilde{v}_j(x) = \frac{\int v_j \pi(x|v)\phi(dv)}{\int \pi(x|v)\phi(dv)} \tag{5}$$

Then it is a two-line matter to verify

Theorem 1. *The mixed density* $\tilde{\pi}(x)$ *is the equilibrium density of the process with infinitesmal generator* $\tilde{T}$.

Despite the formal immediacy of the result, there are substantial and interesting implications. First, note that we have extended the family of processes F with generator (1) for varying v to a family $\tilde{F}$ with generator (4) for varying mixing measure ϕ. The theorem states that the equilibrium density for a member of this extended class of processes is just the mixed density $\tilde{\pi}$.

One can view ϕ as a prior distribution for the parameter v, and $\tilde{v}$ as a corresponding Bayesian estimate of v based upon the assumption that the fixed-v process has reached equilibrium and that one has observed the current state value x. (The estimate might more appropriately be termed a 'forgetful Bayesian' estimate in that one utilises only the current state value and not previously observed values. This is the very point we take up in later sections.) The effect of modifying v to $\tilde{v}$ is to introduce a feedback into the process, by which a transition rule of form (1) is adopted, but the rate vector is adapted to that value which seems most concordant with the current value of x. One might conjecture that the effect of this feedback is to perpetuate the status quo; to adapt the transition rule so that it will tend to yield as outcome the very state value which is current.

This effect, if it exists, should be particularly marked in the case when ϕ gives mass only to a discrete set of values $v^{(r)}$, so that

$$\tilde{\pi}(x) = \sum_r \phi_r \pi(x|v^{(r)}) \tag{6}$$

say. State space will then decompose itself into sets D_r, where D_r is the set in which $\tilde{\nu}(x) \sim \nu^{(r)}$. (More correctly, D_r is the set in which the maximum likelihood estimate $\hat{\nu}(x)$ equals $\nu^{(r)}$. The forgetful Bayesian estimate $\tilde{\nu}(x)$ supplies a smoothed version of $\hat{\nu}(x)$.) The 'perpetuation of the status quo' conjecture would then amount more specifically to the conjecture that the D_r constitute basins of attraction in state space, escape from any given D_r being rendered more difficult by the presence of feedback. The rth term in (6) could then presumably be identified as the component of distribution corresponding to sojourn in D_r.

One may now ask whether the 'forgetful Bayesian' estimate $\tilde{\nu}$ might not be converted into a genuine Bayesian estimate based upon a history rather than a moment of observation (supposedly, for a fixed-ν process in equilibrium). This was John Bather's query. One imagines that the effect of such an increased dependence upon past history would be to increase attachment to the status quo. In view of this natural conjecture, the results below are interesting.

2. Incorporation of State-History in the Construction. Rather than assuming a complete history of observation we shall assume the state variable augmented by variables which store statistical information from past observations. Suppose that the state variable x is supplemented by a variable y such that (i) y(t) is a deterministic function of current x-history $\{x(s); s \leq t\}$ and (ii) the process {x.y} is Markov. This implies that y must obey a deterministic Markov up-dating rule, so that $y(t+\Delta)$ is a function of y(t) and $\{x(s); t < s \leq t+\Delta\}$ for any $\Delta > 0$. For example, y(t) might be a partial history: $y(t) = \{x(s); t\text{-}h < s \leq t\}$. A more manageable choice would be that in which y was a finite-dimensional vector satisfying a differential equation

$$\dot{y} = a(x, y) \tag{7}$$

One could imagine a scheme of this kind for which y(t) supplied quite a good summary of those aspects of the history of x relevant for estimation of ν; we give an example in the next section.

We shall now apply the ideas of the last section to the Markov process with state variable (x, y). The infinitesmal generator defined previously by (1) will now be modified so that it has the action

$$T^* (\nu) = \Sigma \nu_j T_j^* + T^o. \tag{8}$$

Here T_j^* induces the transitions in (x, y) corresponding to the transitions in x induced by T_j, and T^0 reflects the changes in y in the absence of any change in x: the purely deterministic change implied by its deterministic Markov evolution rule.

The (x, y) process will have an equilibrium density $\pi(x, y|\nu)$ relative to an appropriate measure. This will have the equilibrium density $\pi(x|\nu)$ of the simple x-process as marginal x-density, because x-dynamics are not affected by the values of y.

We can now introduce the mixed density

$$\tilde{\pi}(x, y) = \int \pi(x, y|\nu)\,\phi(d\nu) \tag{9}$$

the 'forgetful Bayesian' estimate

$$\tilde{\nu}(x, y) = \frac{\int \nu\pi(x, y|\nu)\,\phi(d\nu)}{\int \pi(x, y|\nu)\,\phi(d\nu)} \tag{10}$$

and the process in which $\tilde{\nu}$ replaces ν in the transition rule (8).

Theorem 2. *The process for which* $\tilde{\nu}(x, y)$ *replaces* ν *in the definition* (8) *of the infinitesmal generator has equilibrium distribution* $\tilde{\pi}(x, y)$. *Furthermore, the marginal* x*-density of this distribution is the density* $\tilde{\pi}(x)$ *defined previously by* (3).

Proof. Since the infinitesmal generator is linear in ν, Theorem 1 is applicable to the compound process $\{x, y\}$ and the first assertion follows. The second assertion follows from (9) and the fact that $\pi(x|\nu)$ is the marginal of $\pi(x, y|\nu)$. □

Thus, we can indeed modify the mixing construction so that $\tilde{\nu}$ is an estimate of ν based upon aspects of x-history as well as upon current x. Surprisingly, this does not affect the equilibrium distribution of the x-process. However, it must affect the transient behaviour. In particular, one would expect it to increase the 'persistence' of the process, and to deepen the basins of attraction D_r. We shall investigate this conjecture by example.

3. An Example. We shall take a linear Gaussian process as example, because this permits explicit analysis. There would be more interest in a case for which the mixing distribution ϕ was discrete; the fact that ν will follow a continuous distribution will mean that there will not be discrete basins of attraction, and 'perpetuation of the status quo' will at best correspond to an increased sluggishness of movement rather than to temporary capture in some subset of state space. However, the case we consider is substantial enough to throw up some surprises.

Suppose that the x-process has a scalar state variable obeying the stochastic differential equation

$$\dot{x} = \alpha(\nu - x) + \varepsilon \tag{11}$$

where $\alpha > 0$ and ε is white noise of power c. The infinitesmal generator has the effect

$$T\psi(x) = \alpha(\nu - x)\psi'(x) + (c/2)\psi''(x)$$

and so is indeed linear in ν. In equilibrium the process has a $N(\nu, (c/2\alpha))$ density; this defines $\pi(x|\nu)$. The parameter ν is then interpretable as the equilibrium mean of x. If we choose the mixing measure ϕ so that ν is $N(0, \tau)$ then we find that

$$\tilde{\nu} = \frac{2\alpha x/c}{(2\alpha/c) + (1/\tau)} = x - x(1 + 2\tau\alpha\ c)^{-1}$$

For the mixed process x is $N(0, (c/2\alpha) + \tau)$ in equilibrium; this defines $\tilde{\pi}(x)$. The effect of feedback is to replace a drift towards ν by a more sluggish drift, ultimately to the origin.

Consider now the supplementation of x by a scalar y obeying the equation

$$\dot{y} = \beta(x - y). \tag{12}$$

where $\beta > 0$. Thus y is an exponentially-damped average of past x, and, for a fixed-ν process, would provide good information on ν. As β becomes infinite the damping becomes infinite and y reduces to x itself. As β tends to zero then y converges to a long-term average of x. It is for small β that one would expect the mixed version of the process to show considerable persistence.

Consider the process {x, y} specified by (11) and (12). The variables x and y are jointly normal when the process has reached equilibrium, both having ν as mean. Calculation shows the spectral density matrix to be

$$f(\omega) = \frac{c}{\omega^2 + \alpha^2}\begin{bmatrix} 1 & \beta/(-i\omega + \beta) \\ \beta/(i\omega + \beta) & \beta^2/(\omega^2 + \beta^2) \end{bmatrix},$$

this implies an equilibrium covariance matrix

$$\operatorname{cov}(x, y) = \frac{c}{2\alpha}\begin{bmatrix} 1 & \beta/(\alpha+\beta) \\ \beta/(\alpha+\beta) & \beta/(\alpha+\beta) \end{bmatrix}.$$

The joint density can then be verified to have the form

$$\pi(x, y \mid \nu) = A(x, y)\exp\left[-\frac{\alpha(\alpha+\beta)}{\beta c}(y-\nu)^2\right]$$

where A(x, y) is independent of ν. Thus, on the basis of observations x and y the variable y is sufficient for inference on the parameter ν. Taking again the N(0, τ) mixing distribution for ν we have then

$$\tilde{\nu}(x, y) = \frac{(2\alpha(\alpha+\beta)\, y/\beta c)}{(2\alpha(\alpha+\beta)/\beta c) + (1/\tau)} = \gamma y \tag{13}$$

say. This converges to y itself as β tends to zero.

For the mixed process relation (11) then becomes modified to

$$\dot{x} = \alpha(\gamma y - x) + \varepsilon \tag{14}$$

this again to be coupled with relation (12). One finds from these relations that the x-process has spectral density function

$$f(\omega) = \frac{c(\omega^2+\beta^2)}{|(i\omega+\alpha)(i\omega+\beta) - \alpha\beta\gamma|^2} = \frac{c(\omega^2+\beta^2)}{(\omega^2+\chi^2)(\omega^2+\eta^2)} \tag{15}$$

say, where

$$\chi + \eta = \alpha + \beta$$

$$\chi\eta = \alpha\beta(1-\gamma) = \alpha\beta\{1 + 2\alpha(\alpha+\beta)\tau/(\beta c)\}^{-1}. \tag{16}$$

The autocovariance of the x-process is then

$$\tau(s) = \frac{c}{2\chi\eta(\chi^2-\eta^2)}[(\chi^2-\beta^2)\eta e^{-\chi s} - (\eta^2-\beta^2)\chi e^{-\eta s}] \tag{17}$$

In particular,

$$\tau(0) = \frac{c(\beta^2+\chi\eta)}{2\chi\eta(\chi+\eta)} = (c/2\alpha) + \tau.$$

This is independent of β, in agreement with the second assertion of Theorem 2.

A measure of persistence might be $-\tau'(0)/\tau(0)$; the rate of decay of the autocorrelation coefficient at zero lag. We find from expression (17) that $\tau'(0) = -c/2$ so that

$$\tau'(0)/\tau(0) = \alpha c(c+2\alpha\tau)^{-1}.$$

This expression decreases with increasing τ, but is quite independent of β. So, by this measure, incorporation of process history into the estimate of ν, and so into the dynamics of the process, has no effect at all upon the degree of persistence of the process.

We can see what effect this incorporation has by examining the autocorrelation function (17). It follows from (16) that one of χ or η become small as β becomes small. Specifically, we can take it that $\chi = \alpha + O(\beta)$ and $\eta = O(\beta)$. Thus, for small β the term $\exp(-\eta s)$ in (17) is slowly decaying. On the other hand, it has a coefficient of order β. One could thus say that, for small β, the variable x has a persistent component (i.e. a component of slowly decaying autocovariance) but that this component is of small amplitude.

Whether this conclusion is valid more generally has yet to be established. What it seems to imply is that the incorporation of a long-term state-average into the dynamics of the process by the construction of section 2 indeed produces a correspondingly persistent component in the process, but one of negligible magnitude, possibly because of the very stochastic stability of this average.

This work was carried out during the author's tenure of a Science and Engineering Research Council Senior Fellowship.

Reference

Whittle, P. (1990). A construction for multi-modal processes, and a potential memory device. To appear *J. Appl. Prob.*, **27**, 146-155.

ACCUMULATION POINTS OF A PARTICULAR NORMALIZED RANDOM WALK

Andre Adler
Illinois Institute of Technology

Abstract

It is shown that a certain hypothesis is unnecessary when obtaining a generalized one-sided Law of the Iterated Logarithm. Also, as in the independent and identically distributed case the [almost certain] set of accumulation points is an interval. It turns out that our set of limit points is the entire interval $[c, \infty]$, for some finite nonzero constant c.

Results. Herein we establish the almost sure set of accumulation points of normalized weighted sums of independent copies of an unbounded asymmetrical random variable. The random variables under examination are of a particularly peculiar type. They permit the existence of a [weighted] Strong Law of Large Numbers, where the limit is not the common expectation (which may not even exist). We will only consider random variables with either mean zero or without finite first absolute moment. For a more extensive survey of previous results see Adler (1989) and Adler (1990). From Adler and Rosalsky (1989) it is clear that the weights $\{a_n, n \geq 1\}$ should satisfy $n|a_n| \uparrow$ and $\Sigma_{k=1}^{n} |a_k| = O(n|a_n|)$. So, as in Adler (1990) we consider weights of the form $a_n = n^{\alpha}$, $\alpha > -1$.

Let $\{X, X_n, n \geq 1\}$ be i.i.d. random variables and $S_n = \Sigma_{k=1}^{n} k^{\alpha} X_k$ denote our partial sums. Define

$$\tilde{\mu}(x) = \int_x^{\infty} P\{|X| > t\}\, dt \quad \text{if } EX = 0$$

and

$$\mu(x) = \int_0^{x} P\{|X| > t\}\, dt \quad \text{when } E|X| = \infty.$$

Our hypotheses are $E(X^-/\tilde{\mu}(X^-)) < \infty$ and $\tilde{\mu}(x) \sim \tilde{\mu}(x \log_2 x)$ when the mean is zero. Similarly we assume that $E(X^-/\mu(X^-)) < \infty$ and $\mu(x) \sim \mu(x \log_2 x)$ when $E|X| = \infty$. (It is customary to define log log x as $\log_2 x$.)

Next, we establish the sequence $\{c_n, n \geq 1\}$ by way of

$$c_n = \begin{cases} n\tilde{\mu}(c_n) & \text{if} \quad EX = 0 \\ n\mu(c_n) & \text{if} \quad E|X| = \infty . \end{cases}$$

Then our norming sequence is obtained via $b_n = n^{\alpha}c_n$. Lastly, purely for simplicity, we choose to call our limit inferior L_{α}. Hence

$$L_{\alpha} = \begin{cases} -(\alpha+1)^{-1} & \text{if} \quad EX = 0 \\ (\alpha+1)^{-1} & \text{if} \quad E|X| = \infty . \end{cases}$$

In generalizing the results of Adler (1990) it is shown that the hypothesis $\tilde{\mu}(b_n) = O(\tilde{\mu}(c_n))$, when $\alpha \,\varepsilon(-1, -1/2]$, is unnecessary.

THEOREM. If $\alpha > -1$, then with probability one

$$\liminf_{n\to\infty} \frac{S_n}{b_n} = L_{\alpha} \quad \text{and} \quad \limsup_{n\to\infty} \frac{S_n}{b_n} = \infty.$$

PROOF. In view of Adler (1990) it suffices to show that

$$\frac{\Sigma_{k=1}^{n} k^{\alpha}[X_k - EXI(|X| \le c_n)]}{b_n} \xrightarrow{P} 0,$$

when $EX = 0$ and $-1 < \alpha \le -1/2$. This will be proved via Adler, Rosalsky and Taylor (1991). Hence we must verify

$$c_n \uparrow, \ \frac{c_n}{n} \downarrow, \sum_{k=1}^{n} k^{\alpha P} = o(b_n^P), \text{and} \sum_{k=1}^{n} \frac{b_k^P}{k^{2+\alpha P}} \cdot \sum_{j=1}^{n} j^{\alpha P} = O(b_n^P),$$

for some $p \in (1, -1/\alpha)$.

Clearly $c_n \uparrow$ and $c_n/n \downarrow$. Let $p \in (1, -1/\alpha)$. Since $\tilde{\mu}(c_n)$ is slowly varying at infinity

$$\sum_{k=1}^{n} k^{\alpha p} \le \frac{(n+1)^{\alpha p+1}}{\alpha p+1} = o([n^{\alpha+1}\tilde{\mu}(c_n)]^P) = o(b_n^P)$$

and

$$\sum_{k=1}^{n} \frac{b_k^p}{k^{2+\alpha p}} \sum_{j=1}^{n} j^{\alpha p} = \sum_{k=1}^{n} \frac{[k^{\alpha+1}\tilde{\mu}(c_k)]^P}{k^{2+\alpha p}} \cdot \sum_{j=1}^{n} j^{\alpha p}$$

$$= \sum_{k=1}^{n} k^{p-2}[\tilde{\mu}(c_k)]^P \cdot \sum_{j=1}^{n} j^{\alpha P}$$

$$= O(n^{p-1}[\tilde{\mu}(c_n)]^p) \cdot O(n^{\alpha p+1})$$

$$= O([n^{\alpha+1}\tilde{\mu}(c_n)]^p) = O(b_n^p).$$

This completes the proof. □

Utilizing techniques similar to those that can be found in Fernholz (1980) it follows from the above theorem that every point in the interval $[L_\alpha, \infty]$ is an almost certain accumulation point of S_n/b_n. This can be accomplished by the verification of a couple of claims. The first is that the hypothesis $\tilde{\mu}(x) \sim \tilde{\mu}(x\log_2 x)$ (or $\mu(x) \sim \mu(x\log_2 x)$) implies that $c_n \sim c_{n+1}$ and hence $b_n \sim b_{n+1}$. Next one shows that $X_n^- = o(c_n)$ a.s. (This follows immediately from Fernholz (1980).) These two together with

$$P\{\frac{S_{n+1}}{b_{n+1}} \le x \le \frac{S_n}{b_n}, \text{ i.o. } (n)\} = 1,$$

for all $x \in (L_\alpha, \infty)$, leads us to the desired conclusion.

References

[1] Adler, A. (1989). Exact sequences for sums of independent random variables. In *Almost Everywhere Convergence* (G. Edgar and L. Sucheston, Ed.), Academic Press, Boston, 11-29.

[2] Adler, A. (1990). Generalized one-sided laws of the iterated logarithm for random variables barely with or without finite mean. *J. Theor. Probab.*, 3, 587-597.

[3] Adler, A. and Rosalsky, A. (1989). On the Chow-Robbins "fair" games problem. *Bull. Inst. Math. Acad. Sinica, 17*, 211-227.

[4] Adler, A., Rosalsky, A. and Taylor, R. L. (1991). A weak law for normed weighted sums of random elements in Rademacher type p Banach spaces. *J. Mult. Analy.* 37.

[5] Fernholz, L. T. (1980). Limit points associated with generalized iterated logarithm laws. *Ann. Probab., 8*, 390-392.

UNIFORM CONVERGENCE OF MARTINGALES IN THE ONE-DIMENSIONAL BRANCHING RANDOM WALK

J.D. Biggins
The University of Sheffield

Abstract

In the supercritical branching random walk an initial person has children whose positions are given by a point process $Z^{(1)}$. Each of these then has children in the same way, with the positions of children in each family, relative to their parent's, being given by independent copies of $Z^{(1)}$, and so on. For any value of its argument, λ, the Laplace transform of the point process of nth generation people, normalized by its expected value, is a martingale, the usual branching process martingale being a special case. Here it is shown that under certain conditions these martingales converge uniformly in λ, almost surely and in mean. A consequence of this result is that the limit is, in an appropriate region, analytic in λ.

1. Introduction. This paper considers the one dimensional supercritical branching random walk. The process starts with a single initial ancestor at the origin. She has children, forming the first generation, with their positions on the real line, R, being given by a point process $Z^{(1)}$. Each of these children then has offspring in a similar way, with the positions of each new family relative to their parent being given by independent copies of $Z^{(1)}$. This gives the point process of second generation individuals, denoted by $Z^{(2)}$. Subsequent generations are formed similarly, yielding $Z^{(n)}$ as the nth generation point process. Let $\{z_r^{(n)} : r\}$ be an enumeration of the positions of the nth generation people.

Let μ by the intensity measure of $Z^{(1)}$ then, as is well known, μ^{n*} (the n-fold convolution of μ) is the intensity measure of $Z^{(n)}$. As the process is supercritical we have $\mu(R) > 1$. Let $m(\lambda)$ be the Laplace transform of μ. Then

$$\begin{aligned} m(\lambda) &= \int e^{-\lambda x}\mu(dx) \\ &= E\int e^{-\lambda x} Z^{(1)}(dx) \\ &= E\sum_r e^{-\lambda z_r^{(1)}} \end{aligned}$$

and hence

$$\begin{aligned} m(\lambda)^n &= \int e^{-\lambda x}\mu^{n*} \\ &= E\int e^{-\lambda x} Z^{(n)}(dx) \\ &= E\sum_r e^{-\lambda z_r^{(n)}}. \end{aligned}$$

We will adopt the convention that the real and imaginary parts of λ are θ and η respectively, so that $\lambda = \theta + i\eta \in \boldsymbol{C}$, where $\boldsymbol{C}$ is the complex numbers.

The description of the process given above implies that, for any set A,

$$Z^{(n+1)}(A) = \sum_r Z_{n,r}^{(1)}(A - z_r^{(n)}), \tag{1.1}$$

where $\{Z_{n,r}^{(1)} : r\}$ are independent copies of $Z^{(1)}$ with $Z_{n,r}^{(1)}$ giving the relative positions of the family of $z_r^{(n)}$. Let $F^{(n)}$ be the σ-field containing all information about the first n generations. Then (1.1) implies that

$$\begin{aligned} E\left(\int e^{-\lambda x} Z^{(n+1)}(dx) \,|\, F^{(n)}\right) &= \sum_r E\left(\int e^{-\lambda x} Z_{n,r}^{(1)}(dx - z_r^{(n)}) \,\middle|\, F^{(n)}\right) \\ &= \sum_r E\left(\int e^{-\lambda(x + z_r^{(n)})} Z_{n,r}^{(1)}(dx) \,|\, F^{(n)}\right) \\ &= \sum_r E\left(\int e^{-\lambda x} Z_{n,r}^{(1)}(dx) \,|\, F^{(n)}\right) e^{-\lambda z_r^{(n)}} \\ &= m(\lambda)\int e^{-\lambda x} Z^{(n)}(dx). \end{aligned}$$

Therefore, if $m(\lambda)$ is finite and non-zero,

$$W^{(n)}(\lambda) = m(\lambda)^{-n}\int e^{-\lambda x} Z^{(n)}(dx)$$

is a martingale with respect to $F^{(n)}$. These martingales and variants of them for similar processes have been considered often, for example, by Watanabe (1967), Joffe et al. (1973), Kingman (1975), Biggins (1977), Wang (1980), Uchiyama (1982) and Neveu (1988).

Notice that $W^{(n)}(\theta)$ is a non-negative martingale and so converges almost surely for all θ. However only when it converges also in mean can we be confident that the limit is not trivial. In particular $W^{(n)}(0)$ is the classical branching process martingale, for which the condition

$$EW^{(1)}(0)\log^{+}(W^{(1)}(0))<\infty$$

is necessary and sufficient for convergence in mean. Let Ω^0 be the interior of $\{\lambda : m(\theta) < \infty\}$; we will assume throughout that Ω^0 is non-empty. Then, if $\theta \in \Omega^0$, it is shown in Biggins (1977) that the conditions

$$EW^{(1)}(\theta)\log^{+}(W^{(1)}(\theta)) < \infty \tag{1.2}$$

and

$$\theta \in \{\theta : -\log(m(\theta)) < -\theta m'(\theta)/m(\theta)\} \tag{1.3}$$

are necessary and sufficient for $EW(\theta) = 1$, and hence for $W^{(n)}(\theta)$ to converge in mean.

The set described by (1.3) is actually an interval and, to avoid complications with end points, we will let T be its intersection with Ω^0, so that T is an open interval. Hence

$$T = \{\theta : \theta \in \Omega^0, -\log(m(\theta)) < -\theta m'(\theta)/m(\theta)\}. \tag{1.4}$$

The main results here concern the convergence of $W^{(n)}(\lambda)$ or $W^{(n)}(\theta)$ as a sequence of functions. Under suitable conditions there is an open set Λ^* in C, containing T, with $W^{(n)}(\lambda)$ converging uniformly on any compact subset of Λ^*, almost surely and in mean. A consequence of this is that the limit, $W(\lambda)$, is actually analytic on Λ^*. Under a rather weaker moment condition, a slightly different approach yields the uniform convergence of $W^{(n)}(\theta)$ to $W(\theta)$ on compact subsets of T, implying that $W(\theta)$ is continuous on T.

Joffe et al. (1973) give a result on the uniform convergence of $W^{(n)}(i\eta)$ for a particular case of the process considered here. They adopt an elegant approach through results on convergence of martingales taking values in a Banach space, with a vital step in the proof being the verification that the limit $W(i\eta)$ is continuous. In contrast here the convergence of $W^{(n)}(\lambda)$ to $W(\lambda)$ will be tackled directly, yielding the convergence of the appropriate Banach space valued martingale and the properties of the limit as consequences.

A more detailed study of the convergence of $W^{(n)}(\lambda)$ to $W(\lambda)$ is contained in Biggins (1989). There a uniform convergence result is obtained for the branching random walk on R^p both in discrete and continuous time. An application of these results is also given, large deviation results for $Z^{(n)}$ being obtained using them.

2. The Main Results. We consider first the convergence of $W^{(n)}(\lambda)$ in a suitable region of C.

Theorem 1. *If for some* $\gamma > 1$

$$EW^{(1)}(\theta)^{\gamma} < \infty \text{ for all } \theta \in T \tag{2.1}$$

and Λ^* *is defined by*

$$\Lambda^* = \{\lambda \in \Omega^0 : m(\alpha\theta)/|m(\lambda)|^{\alpha} < 1 \text{ for some } \alpha \in (1,\gamma]\}$$

then $W^{(n)}(\lambda)$ *converges uniformly, to* $W(\lambda)$, *on any compact subset of* Λ^*, *almost surely and in mean.*

It is not too hard to show that

$$\Lambda^* = \bigcup_{1<\alpha\leq\gamma} \operatorname{int}\{\lambda \in \Omega^0 : m(\alpha\theta)/|m(\lambda)|^{\alpha} < 1\} \tag{2.2}$$

so Λ^* is open. Furthermore computing the derivative of $m(\alpha\theta)/m(\theta)^{\alpha}$ with respect to α and setting $\alpha = 1$ confirms that the set T, defined at (1.4), is the intersection of Λ^* with the real axis.

If F is a compact subset of Λ^* the assertion of the theorem is that

$$\sup\{\ |W^{(N)}(\lambda) - W^{(n)}(\lambda)| : \lambda \in F, N \geq n\ \}$$

converges to zero almost surely and in mean as $n \to \infty$. It will then follow that $W(\lambda)$ exists on F and that

$$\sup\{|W^{(n)}(\lambda) - W(\lambda)| : \lambda \in F\}$$

also converges to zero in both senses.

The approach we will take to this result relies heavily on Cauchy's integral formula and so does not lend itself to considering convergence for $\theta \in T$ alone.

Furthermore, with this approach, I can see no way to escape from the rather strong moment condition (2.1). A weaker moment condition suffices in considering the convergence of $W^{(n)}(\theta)$ on T to which we now turn.

Theorem 2. *If*

$$EW^{(1)}(\theta)\,(\log^+(W^{(1)}(\theta)))^{3/2} < \infty \text{ for all } \theta \in T \tag{2.3}$$

then $W^{(n)}(\theta)$ *converges uniformly to* $W(\theta)$ *on compact subsets of* T*, almost surely and in mean.*

The two moment conditions (2.1) and (2.3) are easily seen to be equivalent to the alternatives resulting if $W^{(1)}(\theta)$ is replaced by $\int e^{-\theta x} Z^{(1)}(dx)$. Now, if g is any convex function, $Eg\left(\int e^{-\theta x} Z^{(1)}(dx)\right)$ is convex in θ and so its domain of finiteness must be an interval. Furthermore it must be uniformly bounded on compact subsets of the interior of this interval. The conditions (2.1) and (2.3) simply insist that, for particular g, this interval of finiteness should include all of T.

Differentiation of $E(W^{(n+1)}(\theta) \mid F^{(n)}) = W^{(n)}(\theta)$ shows that $W^{(n)}(\theta)'$ is also a martingale, as are all higher derivatives. The proof of Theorem 2 yields, in the course of its proof, the following result about this martingale.

Theorem 3. *If* $\theta \in T$ *with*

$$EW^{(1)}(\theta)\,(\log^+(W^{(1)}(\theta)))^{3/2} < \infty \tag{2.4}$$

and

$$E|W^{(1)}(\theta)'| \log^+(|W^{(1)}(\theta)'|) < \infty \tag{2.5}$$

then $W^{(n)}(\theta)'$ *converges almost surely and in mean.*

At the end of section 4 it is indicated how the method of proof extends to higher derivatives, at some notational expense, but the details are not considered here. It is perhaps worth noting that in proving Theorem 2 we will show that the moment condition (2.3) implies (2.4) and (2.5). Of course under the stronger moment condition (2.1) Theorem 1 holds, and then all derivatives converge.

The proof of the main results relies heavily on the following lemma which is proved in Biggins (1989).

Lemma 1. *If* $\{X_r\}$ *are independent complex random variables with* $E(X_r) = 0$ *or, more generally, martingale differences, then*

$$E|\Sigma X_r|^\alpha \le 2^\alpha \Sigma E|X_r|^\alpha$$

for $1 \le \alpha \le 2$.

This, together with Jensen's inequality, immediately yields the following result for the branching random walk.

Lemma 2. *If, given* $F^{(n)}$, $\{X_r\}$ *are independent identically distributed copies of* X *with* $EX = 0$ *then*

$$E\left|\sum_r \frac{e^{-\lambda z_r^{(n)}}}{m(\lambda)^n} X_r\right| \le 2\left(\frac{m(\alpha\theta)^n}{|m(\lambda)|^{\alpha n}} E|X|^\alpha\right)^{1/\alpha}$$

for $1 \le \alpha \le 2$.

3. Proof of Theorem 1. It will suffice to show that for each $\lambda_0 \in \Lambda^*$ uniform convergence holds in a disc centred at λ_0 for then a standard covering argument completes the proof. Denote the disc of radius ρ centred at λ_0 by $D_{\lambda_0}(\rho)$.

Given $\lambda_0 \in \Lambda^*$, we can, because of the representation (2.2), find $\alpha \in (1, \gamma]$ and ρ such that

$$D_{\lambda_0}(3\rho) \subset \{\lambda \in \Omega^0 : m(\alpha\theta) / |m(\lambda)|^\alpha < 1\}. \tag{3.1}$$

We will demonstrate convergence on $D_{\lambda_0}(\rho)$.

Let Γ be the boundary of $D_{\lambda_0}(2\rho)$ and suppose f is analytic on $D_{\lambda_0}(3\rho)$, then, writing

$$\Gamma = \{z(t) : z(t) = \lambda_0 + 2\rho e^{2\pi i t},\ t \in [0, 1)\}, \tag{3.2}$$

Cauchy's integral formula gives

$$f(\zeta) = \frac{1}{2\pi i}\int_\Gamma \frac{f(z)}{z - \zeta} dz$$

$$= \int_0^1 \frac{f(z(t))}{z(t) - \zeta} 2\rho e^{2\pi i t} dt.$$

Therefore

$$\sup\{|f(\zeta)| : \zeta \in D_{\lambda_0}(\rho)\} \le 2\int_0^1 |f(z(t))|\, dt, \tag{3.3}$$

where $z(t)$ parameterises the boundary of Γ and is defined at (3.2).

As $W^{(N)}(\zeta) - W^{(n)}(\zeta)$ is an analytic function on Λ^*, (3.3) yields immediately that

$$\sup\{|W^{(N)}(\zeta) - W^{(n)}(\zeta)| \ : \zeta \in D_{\lambda_0}(\rho)\}$$

$$\leq 2\int_0^1 |W^{(N)} - W^{(n)}|\, dt$$

$$\leq 2\int_0^1 \sum_{r=n}^{N-1} |W^{(r+1)} - W^{(r)}|\, dt$$

$$\leq 2\int_0^1 \sum_{r=n}^{\infty} |W^{(r+1)} - W^{(r)}|\, dt, \tag{3.4}$$

where we have suppressed $z(t)$ in the integrand. The proof will be complete if we show that (3.4) converges to zero almost surely and in mean as $n \to \infty$, and for this it suffices to verify that it has finite expectation. Now note that

$$E\int_0^1 \sum_{n=0}^{\infty} |W^{(n+1)} - W^{(n)}|\, dt$$

$$= \int_0^1 \sum_{n=0}^{\infty} E|W^{(n+1)} - W^{(n)}|\, dt$$

$$\leq \sup\{\sum_n E|W^{(n+1)}(\lambda) - W^{(n)}(\lambda)| : \lambda \in \Gamma\ \}. \tag{3.5}$$

Furthermore, with the obvious extension of the notation of (1.1),

$$W^{(n+1)}(\lambda) - W^{(n)}(\lambda) = \sum_r \frac{e^{-\lambda z_r^{(n)}}}{m(\lambda)^n}\left\{W_{n,r}^{(1)}(\lambda) - 1\right\} \tag{3.6}$$

and so Lemma 2 can be applied to estimate (3.5).

Observe first that

$$E|W^{(1)}(\lambda) - 1|^{\alpha} \leq 2^{\alpha}(E|W^{(1)}(\lambda)|^{\alpha} + 1)$$

$$\leq 2^{\alpha}\left(\frac{m(\theta)^{\alpha}}{|m(\lambda)|^{\alpha}} EW^{(1)}(\theta)^{\alpha} + 1\right)$$

and so, by virtue of the moment condition (2.1) and the remarks immediately after Theorem 2, is uniformly bounded for $\lambda \in \Gamma$. Hence, applying Lemma 2,

$$\sum_n E\left|W^{(n+1)}(\lambda) - W^{(n)}(\lambda)\right| \le K \sum_n \left(\frac{m(\alpha\theta)}{|m(\lambda)|^\alpha}\right)^{n/\alpha}$$

for $\lambda \in \Gamma$ but, because of (3.1) and (3.2),

$$\sup\left\{\frac{m(\alpha\theta)}{|m(\lambda)|^\alpha} : \lambda \in \Gamma\right\} < 1$$

and so (3.5) is indeed finite.

4. Proof of Theorems 2 and 3. Just as in the proof of Theorem 1 it will suffice to show that for each $\theta_0 \in T$ uniform convergence holds in an interval around θ_0. As $\theta_0 \in T$, there is a $\beta > 1$ with

$$\frac{m(\beta\theta_0)}{m(\theta_0)^\beta} < 1$$

so we may choose ρ sufficiently small that

$$(\theta_0 - 2\rho, \theta_0 + 2\rho) \subset \left\{\theta : \frac{m(\beta\theta)}{m(\theta)^\beta} < 1\right\} \tag{4.1}$$

(which implies that $\theta_0 \pm 2\rho \in T$). Now let $D = (\theta_0 - \rho, \theta_0 + \rho)$. Then

$$\delta = \sup\left\{\frac{m(\beta\theta)^{1/\beta}}{m(\theta)} : \theta \in D\right\} < 1. \tag{4.2}$$

In place of the bound (3.3), based on Cauchy's integral formula, we employ the following simple estimate

$$\sup\{|f(\theta)| : \theta \in D\} \le f(\theta_0) + \int_D |f'(\zeta)|\, d\zeta.$$

Hence

$$\sup\left\{\left|W^{(N)}(\theta) - W^{(n)}(\theta)\right| : \theta \in D\right\}$$

$$\le \left|W^{(N)}(\theta_0) - W^{(n)}(\theta_0)\right| + \int_D \left|(W^{(N)}(\theta) - W^{(n)}(\theta))'\right| d\theta$$

$$\leq \sum_{r=n}^{\infty} \left|W^{(r+1)}(\theta_0) - W^{(r)}(\theta_0)\right|$$

$$+ \sum_{r=n}^{\infty} \int_D \left|(W^{(r+1)}(\theta) - W^{(r)}(\theta'))'\right| d\theta \tag{4.3}$$

and the result will be proved by showing that the final bound here has finite expectation. The second term on the right is the harder one to deal with so our discussion focuses on it. In considering it Theorem 3 will be proved as a by-product. Much as in (3.5), it will be enough to show that

$$\sup\{\sum_n E\left|(W^{(n+1)}(\theta) - W^{(n)}(\theta))'\right| : \theta \in D\} < \infty.$$

Differentiation of (3.6) gives

$$(W^{(n+1)}(\theta) - W^{(n)}(\theta))'$$

$$= \sum_r \frac{e^{-\theta z_r^{(n)}}}{m(\theta)^n} \left\{\left(-z_r^{(n)} - \frac{nm'(\theta)}{m(\theta)}\right)(W_{n,r}^{(1)}(\theta) - 1) + W_{n,r}^{(1)}(\theta)'\right\} \tag{4.4}$$

Notice that these are the differences of the martingale $\{W^{(n)}(\theta)'\}$ and, under suitable moment conditions, Lemma 1 would apply to (4.4) to give a bound on the expectation of its absolute value. When the resulting bounds have a finite sum (over n) this yields the convergence almost surely and in mean of the martingale. However to allow weaker moment conditions we will use a truncation technique and so need variants of Lemmas 1 and 2, which we now discuss. In these lemmas $\{I_r\}$ will be indicator functions with $I_r^c = 1 - I_r$. These will be used to isolate cases where $|X_r|$ is big.

Lemma 3. *Let $\{X_r, I_r\}$ be a sequence of random vectors which are, given G, independent with $E(X_r \mid G) = 0$ and let N be a (possibly infinite) G measurable random integer then*

$$E\left|\sum^N X_r\right| \leq 2E\sum^N E(|X_r I_r| \mid G) + 2^3\left(E\sum^N E\left(\left|X_r I_r^c\right|^\alpha \mid G\right)\right)^{1/\alpha}$$

for $1 \leq \alpha \leq 2$.

Proof. As $E(X_r \mid G) = 0$

$$E(X_rI_r^c|\,G) = -E(X_rI_r|\,G)\,,$$

consequently

$$\Sigma X_r = \Sigma X_rI_r + \Sigma E(X_rI_r|\,G) + \Sigma(X_rI_r^c - E(X_rI_r^c|\,G))$$

, (4.5)

where, of course, the N has been suppressed in the notation. Now

$$E|\Sigma X_rI_r| \le E\Sigma E(|X_rI_r|\,|G) \tag{4.6}$$

and

$$E|\Sigma E(X_rI_r|\,G)| \le E\Sigma E(|X_rI_r|\,|G)\,. \tag{4.7}$$

The final term on the right of (4.5) requires more work. Note first that for any random variable X and $\alpha \ge 1$

$$\begin{aligned} E|X - E(X)|^\alpha &\le 2^\alpha(E|X|^\alpha + |E(X)|^\alpha) \\ &\le 2^{\alpha+1}E|X|^\alpha\,. \end{aligned}$$

Now applying Jensen's inequality, Lemma 1 and the inequality just derived we see that

$$\begin{aligned} E\left|\sum(X_rI_r^c - E(X_rI_r^c|\,G))\right| &\le \left(E\left|\sum(X_rI_r^c - E(X_rI_r^c|\,G))\right|^\alpha\right)^{1/\alpha} \\ &= \left(EE\left|\sum(X_rI_r^c - E(X_rI_r^c|\,G))\right|^\alpha\middle|G\right)^{1/\alpha} \\ &\le \left(E2^\alpha\sum 2^{\alpha+1}E\left(\left|X_rI_r^c\right|^\alpha\middle|G\right)\right)^{1/\alpha} \\ &\le 2^3\left(E\sum E\left(\left|X_rI_r^c\right|^\alpha\middle|G\right)\right)^{1/\alpha} \end{aligned} \tag{4.8}$$

for $1 \le \alpha \le 2$. Combining the bounds (4.6), (4.7) and (4.8) with (4.5) completes the proof.

It is worth stating the following special case of Lemma 3 which has conditions appropriate to our context.

Lemma 4. *If, given G, $\{X_r, I_r\}$ are independent identically distributed copies of $\{X, I\}$ with $EX = 0$, and $\{C_r\}$ are G measurable, then*

$$E\left|\sum C_rX_r\right| \le 2E|XI|E\sum|C_r| + 2^3(E|XI^c|^\alpha)^{1/\alpha}(E\sum|C_r|^\alpha)^{1/\alpha}. \tag{4.9}$$

We now return to consideration of (4.4), applying the lemma just obtained to its two parts separately. Consider first

$$E\left|\sum_r \frac{e^{-\theta z_r^{(n)}}}{m(\theta)^n}\left(-z_r^{(n)} - \frac{nm'(\theta)}{m(\theta)}\right)(W_{n,r}^{(1)}(\theta) - 1)\right|, \tag{4.10}$$

which is the more complicated of the two. Obviously we take $F^{(n)}$ as G. Let

$$C_r = \frac{e^{-\theta z_r^{(n)}}}{m(\theta)^n}\left(-z_r^{(n)} - \frac{nm'(\theta)}{m(\theta)}\right),$$

$$X_r = W_{n,r}^{(1)}(\theta) - 1$$

and for the indicator variables let

$$I_r = I\left\{1 + W_{n,r}^{(1)}(\theta) > c^n\right\} \tag{4.11}$$

where $c > 1$ will be fixed later. Let F_θ be the probability measure of the random variable in this indicator; we will use this in bounding the terms obtained in applying Lemma 4. Specifically, observe that

$$E|XI| \le \int_{c^n}^{\infty} xF_\theta(dx) \tag{4.12}$$

and

$$E|XI^c|^\alpha \le \int_1^{c^n} x^\alpha F_\theta(dx) \tag{4.13}$$

To simplify the expressions resulting from the calculation of the other components of (4.9) let the probability measure μ_θ, which has mean zero, be given by

$$\mu_\theta(dx) = \frac{e^{-\theta x}}{m(\theta)}\mu\left(dx + \frac{m'(\theta)}{m(\theta)}\right).$$

Then it is straightforward to check that

$$E\sum_r \frac{e^{-\theta z_r^{(n)}}}{m(\theta)^n} f\left(z_r^{(n)} + \frac{nm'(\theta)}{m(\theta)}\right) = \int f(x)\,\mu_\theta^{n*}\,(dx),$$

and hence

$$E\sum_r \frac{e^{-\alpha\theta z_r^{(n)}}}{m(\alpha\theta)^n} f\left(z_r^{(n)} + \frac{nm'(\theta)}{m(\theta)}\right) = \int f(x + nk(\alpha,\theta))\,\mu_{\alpha\theta}^{n*}\,(dx),$$

where

$$k(\alpha, \theta) = \frac{m'(\alpha\theta)}{m(\alpha\theta)} - \frac{m'(\theta)}{m(\theta)}.$$

Therefore

$$E\sum |C_r| \leq \int |x| \mu_\theta^{n*}(dx) \tag{4.14}$$

and

$$E\sum |C_r|^\alpha \leq \left(\frac{m(\theta\alpha)}{m(\theta)^\alpha}\right)^n \int |x + nk(\alpha, \theta)|^\alpha \mu_{\alpha\theta}^{n*}(dx). \tag{4.15}$$

Applying Lemma 4 to (4.10) using the above calculations will still result in rather unwieldy expressions, so we now restrict attention to $\theta \in D$. Note first that, as $\int x^2 \mu_\theta(dx)$ is continuous on Ω^0 and hence uniformly bounded on D,

$$\begin{aligned}(\int |x|^\alpha \mu_\theta^{n*}(dx))^{1/\alpha} &\leq (\int x^2 \mu_\theta^{n*}(dx))^{1/2} \\ &= n^{1/2}(\int x^2 \mu_\theta(dx))^{1/2} \\ &\leq Kn^{1/2},\end{aligned}$$

for $1 \leq \alpha \leq 2$. (We will use K for a generic constant independent of $\theta \in D$.) Now we take $\alpha = \beta$, so that (4.1) holds. Hence

$$\sup\{k(\beta, \theta) : \theta \in D\} < \infty$$

so that

$$\int |x + nk(\beta, \theta)|^\beta \mu_{\beta\theta}^{n*}(dx) \leq Kn^\beta,$$

and furthermore (4.2) holds. With these estimates and the bounds (4.12), (4.13), (4.14), and (4.15) we can now apply Lemma 4 to see that (4.10) is bounded by

$$K\left(n^{1/2}\int_{c^n}^\infty xF_\theta(dx) + n\delta^n\left(\int_1^{c^n} x^\beta F_\theta(dx)\right)^{1/\beta}\right)$$

which is in turn bounded by

$$.K(n^{1/2}\int_{cn}^\infty xF_\theta(dx) + n\delta^n c^{n(\beta-1)/\beta}(\int_1^\infty xF_\theta(dx))^{1/\beta}) \tag{4.16}$$

Now we choose $c > 1$ but small enough that $\delta c^{(\beta-1)/\beta} < 1$. We now want to show that (4.16) is summable over n, and for Theorem 2 this must be so uniformly in $\theta \in D$. The sum of the second term here will be (uniformly) bounded provided that $\int_1^\infty xF_\theta(dx)$ is. Turning to the first term

$$\sum_n n^{1/2}\int_{c^n}^\infty xF_\theta(dx) = \int_1^\infty x\sum_n n^{1/2}I\left(n \le \frac{\log x}{\log c}\right)F_\theta(dx)$$

$$\le K\int_1^\infty x(\log x)^{3/2}F_\theta(dx) \quad . \tag{4.17}$$

Hence the moment condition (2.4) (or (2.3)) ensures that both sums are finite.

Consideration of the second part of (4.4),

$$E\left|\sum_r \frac{e^{-\theta z_r^{(n)}}}{m(\theta)^n} W_{n,r}^{(1)}(\theta)'\right|,$$

is similar, the indicator variables now being

$$I_r = I\left\{1 + \left|W_{n,r}^{(1)}(\theta)'\right| > c^n\right\}$$

with F_θ^1 as the associated probability measure. The calculation analogous to (4.17) no longer involves $n^{1/2}$ and so leads to the moment condition (2.5). This completes the proof of Theorem 3.

For the proof of Theorem 2 we must show in addition that

$$\sup\left\{\int x(\log x)^{3/2}F_\theta(dx) : \theta \in D\right\} < \infty \tag{4.18}$$

and

$$\sup\left\{\int x(\log x)F_\theta^1(dx) : \theta \in D\right\} < \infty \tag{4.19}$$

To do this let $\theta_1 = \theta_0 - 2\rho$ and $\theta_2 = \theta_0 + 2\rho$; recall that ρ was chosen so that θ_1 and θ_2 are in T. Note that

$$\sup\left\{\frac{(1+|x|)e^{-\theta x}}{e^{-\theta_1 x} + e^{-\theta_2 x}} : \theta \in D\right\} < \infty \tag{4.20}$$

and so, for $\theta \in D$,

$$
\begin{aligned}
W^{(1)}(\theta)+\left|W^{(1)}(\theta)\right| &\leq K\sum_r (1+|z_r^{(1)}|)e^{-\theta z_r^{(1)}} \\
&\leq K\sum_r \left(e^{-\theta_1 z_r^{(1)}}+e^{-\theta_2 z_r^{(1)}}\right) \\
&\leq K(W^{(1)}(\theta_1)+W^{(1)}(\theta_2)) \\
&\leq K\max\left\{W^{(1)}(\theta_1),W^{(1)}(\theta_2)\right\}; \qquad (4.21)
\end{aligned}
$$

hence (2.3) does indeed guarantee (4.18) and (4.19).

This shows that the second term on the right of (4.3) has finite expectation. A similar, but much more straightforward, analysis shows the finiteness of the expectation of the first term. (The analysis also establishes that, when $\theta \in T$, (1.2) is indeed sufficient for $EW(\theta) = 1$.) This completes the proof of Theorem 2.

If we considered higher derivatives of $W^{(n)}(\theta)$ the analogue of (4.4) would now involve each of the derivatives of $W^{(1)}_{n,r}(\theta)$ up to the degree in question. These can be analysed as the components of (4.4) were. However the powers of n multiplying the two parts of (4.16) will depend on which term in the analogue of (4.4) we are considering. The argument dealing with the second term in (4.16) is unaffected by any higher powers of n but they do change the power of $\log x$ appearing in (4.17). It is worth noting too that an estimate like (4.20) will still work if higher powers of $|x|$ are included. Hence a bound like (4.21) also holds for higher derivatives. Combining these considerations we see that analogues of Theorems 2 and 3 can be obtained for higher derivatives by suitably strengthening the moment conditions.

References

[1] Biggins, J.D. (1977). Martingale convergence in the branching random walk. *J. Appl. Prob.* **14** , 25-37.

[2] Biggins, J.D. (1989). Uniform convergence of martingales in the branching random walk. (to appear in *Ann. Probab.*)

[3] Joffe, A. Le Cam, L. and Neveu, J. (1973). Sur la loi des grandes nombres pour les variables aleatories de Bernoulli attacheesaun arbre dyadique. *C. R. Acad. Sc. Paris*, **277A** , 963-964.

[4] Kingman, J.F.C. (1975). The first birth problem for an age-dependent branching process. *Ann. Probab.* **3**, 790-801.

[5] Neveu, J. (1988). Multiplicative martingales for spatial branching processes. In *Seminar on Stochastic Processes, 1987*, eds: E. Çinlar, K.L. Chung, R.K. Getoor. Progress in Probability and Statistics, **15**, 223-241. Birkhäuser, Boston.

[6] Uchiyama, K. (1982). Spatial growth of a branching process of particles living in R^d. *Ann. Probab.* **10**, 898-920.

[7] Wang, F.J.S. (1980). The convergence of a branching brownian motion used as a model describing the spread of an epidemic. *J. Appl. Prob.* **17**, 301-312.

[8] Watanabe, S. (1967). Limit theorem for a class of branching processes. In *Markov Processes and Potential Theory*, ed: J. Chover. Wiley, New York. 205-232.

AN URN MODEL AND THE COALESCENT IN NEUTRAL INFINITE-ALLELES GENETIC PROCESSES

David Branson
University of Essex

Abstract

An urn model is proposed as a useful computational device for obtaining results within the context of infinite-alleles genetic processes incorporating selective neutrality. The model is based on Kingman's coalescent, as developed by Watterson and by Donnelly and Tavaré, and mimics the procedure of tracing a sample's ancestry backwards in time, noting the appearance of common ancestors or new mutants. The structure of the model is possessed by a time-inhomogeneous linear birth-and-death model with immigration, by Moran's model and by a class of other models including the Wright-Fisher model in the limit of large population size. Elementary combinatorial arguments connected with the partial or complete emptying of the urn give rise to a number of results which can be interpreted in terms of the allelic composition of genetic samples and populations. The relationship to other urn models is discussed.

1. Introduction. Some years ago Kingman (1982 a,b,c) introduced the fruitful concept of the coalescent in which a sample of genes is taken from a population and its ancestry traced back in time, noting where there are common ancestors, until one reaches a single common ancestor. This idea has been developed by others, in particular by Watterson (1984) and by Donnelly and Tavaré (1986) who consider in detail the consequences of mutation. In tracing the ancestry backwards we may come to a mutant which introduced a new allele into the population and which was therefore the originating ancestor for a family represented in the sample, all members of the family carrying the same allele. We assume mutation is non-recurrent, that is each mutation produces a previously unknown allele and we assume that selection is absent. We further suppose that the organism we are considering is a monoecious haploid reproducing either asexually or by self-fertilization, so that each individual has just one parent and, in the absence of mutation, the genetic composition of an individual is identical to that of its parent.

In Section 2 we describe an urn model which mimics the coalescent and which keeps track of the time-ordering of the various mutant and non-mutant births, should this be of interest. The model gives a method of constructing the jump chain of Donnelly and Tavaré's coalescent with ages and is put forward as a way of visualizing the process and as a useful computational device for deriving results (particularly when the urn is only partly emptied) which can be interpreted in terms of a number of population models. This is illustrated in Section 4 where elementary combinatorial arguments provide a simple derivation of several results most of which have appeared elsewhere in various contexts.

Donnelly and Tavaré show that their model (and hence our urn model) describes Moran's model and a class of other models with non-overlapping generations including the Wright-Fisher model in the limit of large population size. Tavaré (1989) has shown that the coalescent with ages reproduces the results of the (time-homogeneous) birth-death-immigration model conditioned on population size. In this case the possibility of mutation within a family is excluded, but the introduction of a new allele into the population is modelled by the immigration of an individual carrying this allele, who then becomes the originating ancestor of a family. We show in Section 3 that the structure of our urn model is also possessed by a time-inhomogeneous birth-death-immigration model provided the birth-per-individual and immigration rates are proportional; we discuss the validity of this assumption. For the Moran and birth-death-immigration models, both the population as a whole and a sample chosen from it have the structure of our urn model.

In Section 5 a brief discussion is given of other urn models in the literature of population genetics. We place our model in the context of these others, and point out its differences from them.

2. An Urn Model. The coalescent takes (without replacement) a sample from a population and traces their ancestry back in time, noting where there are common ancestors. We allow the possibility of mutation (or immigration, but for convenience we shall use the word mutation to refer to either sort of event), so in tracing ancestry backwards we may come to a mutant who was the originating ancestor for a family represented in the sample; we trace this branch of the population back no further. That is, we are not concerned here with the prior history of an individual who introduced a previously unknown allele into the population.

We call the pattern obtained by tracing the sample back through the generations to the original ancestor(s) the sample-forest, as it will consist of a collection of trees each headed by a new mutant. We take no notice of a non-mutant birth with the simultaneous death of the parent (such as is allowed in the Moran model) since this leads neither to a node in the sample-forest, nor to the termination of a tree - in this case we simply identify parent and offspring; neither are we interested here in anything outside the sample-forest. In fact in tracing the sample-forest backwards in time we take account of only two categories of event:

(i) coalescence: two branches in the sample-forest merge at a birth (with no mutation) either because two individuals belonging to the forest at a particular time are a parent-offspring pair, or because, in a model where individuals do not survive the birth of their offspring, two members of the sample-forest in one generation have the same parent in the previous generation. In order to unify our treatment we find it helpful in the latter case to pretend that one of the two individuals in the later generation is a surviving parent and the other is its offspring. This is clearly a harmless fiction since parent and offspring are genetically identical.

(ii) mutation: one branch in the sample-forest ends with the birth of a mutant.

In Figure 1 the thick lines show the sample-forest arising from the six individuals denoted by black squares. The sample-forest contains three coalescences and three mutations; it consists of three trees (families) containing one, two and three members. The members of each family are genetically identical, but the families are genetically distinct.

The probability of simultaneous events of type (i) or (ii) is taken to be zero and all events are assumed to be independent. We shall often refer to "survivors" at a particular time: by this we mean the branches of the sample-forest which, in tracing backwards to the time in question, have not been removed by coalescence or mutation. Each event reduces by one the number of survivors, so that, if the original sample was of size m, the forest ends after m events.

We now describe an urn model whose relevance to these ideas will be detailed in the following section. Suppose an urn originally contains m balls of the same size but each of a different colour (not black) representing a sample of size m taken from a population. We remove the balls from the urn in m stages, which we label successively as stage m, stage $m-1$, ..., stage 1 (so that stage i commences with i balls in the urn). In stage m we remove one ball chosen at random and designate it an "offspring" ball. We replace it with a black ball whose size is such that the relative probabilities of picking the black ball or any particular coloured ball are in the ratio θ:1. We now remove a second ball: if it is black, stage m ends (the interpretation will be that the "offspring" was a mutant); if it is coloured, we designate it a "parent" ball, replace it in the urn, remove the black ball, but recolour the "offspring" ball with the colour of its "parent" (this will be interpreted as a coalescence as described above). This completes stage m.

We continue in a similar fashion. In stage i, the urn contains i balls, and the probability that any particular ball is chosen as "offspring", followed by the black ball is

$$\frac{1}{i}\frac{\theta}{\theta+i-1} \quad . \tag{2.1}$$

On the other hand, the probability that any particular ball is chosen as "offspring" and any other particular ball as "parent" is

$$\frac{1}{i}\frac{1}{\theta+i-1}. \tag{2.2}$$

If any stage results in a coalescence we recolour with the "parent's" colour not only the "offspring" at that stage but also all previously removed balls of the same colour (which are "offspring" of "offspring" ...). For example, at stage m we may remove the blue ball as offspring and the red ball as parent, so we replace the red ball but change the colour of the offspring ball from blue to red. At a later stage we may remove the red ball as offspring and the green ball as parent. We replace the green ball but change to green the colour of both balls that are now red (that is not only green's "child" but also its "grandchild" originally coloured blue). Thus at any stage we have painted with the same colour a parent (in the urn) and all its progeny of succeeding generations (outside the urn).

After stage 1 the urn will be empty. Outside the urn will be m balls of l (say) different colours if the black ball was chosen on l occasions. The balls carrying the same colour represent a single family descended from an originating mutant ancestor. That is, in genetic terms, the sample contains l distinct alleles.

3. Population Models. The urn model described in Section 2 will apply to any population model whose transitions from one stage to the next are independent with probabilities given by (2.1) and (2.2). Donnelly and Tavaré (1986), in discussing their coalescent with ages, show that this is the case for Moran's model if we put

$$\theta = Mu/(1-u), \tag{3.1}$$

where M is the population size and u is the probability that any offspring is a mutant. They show it is also the case for a class of models with non-overlapping generations including the large population limit of the Wright-Fisher model for which one lets M tend to infinity and u tend to zero with θ=2Mu finite and nonzero. (Note that their expressions equivalent to (2.2) contain a factor 2 as, following a birth, they do not distinguish between offspring and parent.) Tavare (1989) has discussed the connection between the coalescent with ages and a time-homogeneous birth-death-immigration model where each immigrant is the source of a new allele in the population. However it is likely that fertility, mortality and immigration will vary in time in response to (perhaps seasonal) variations in the environment. Let us therefore suppose that the birth-per-individual, death-per-individual and immigration rates at time t are given respectively by $\lambda(t)$, $\mu(t)$ and $\nu(t)$. The functions $\lambda(t)$ and $\mu(t)$ are arbitrary functions of time, but the same functions for different families; that is, the different alleles possessed by different families confer no advantage or disadvantage concerning birth or death rates. In other words, our model incorporates selective neutrality. For convenience, after a birth has occurred, we arbitrarily label one of the two resulting identical individuals as "parent" and the other as "offspring". Denote the population size at time t by M(t).

We take (without replacement) a sample of size m, and trace its ancestry back in time, as described in Section 2, taking note of coalescences and mutations (immigrations). Suppose stage i+1 has been completed, leaving i survivors, and suppose that a birth or immigration in the population occurs at time τ, decreasing the population size from $M(\tau)$ to $M(\tau)-1$. The probabilities of this event's being an immigration or a birth are, respectively,

$$\nu(\tau)/[\nu(\tau)+(M(\tau)-1)\lambda(\tau)], \quad (M(\tau)-1)\lambda(\tau)/[\nu(\tau)+(M(\tau)-1)\lambda(\tau)].$$

Since each of the M individuals in the population is equally likely to be the new arrival, and each individual is equally likely to be the parent it follows that the probability that any particular individual is a new immigrant is

$$\frac{\nu}{M[\nu+(M-1)\lambda]} \tag{3.2}$$

whereas the probability that any two particular individuals are respectively a new-born individual and its parent is

$$\frac{(M-1)\lambda}{M(M-1)\,[\nu+(M-1)\lambda]}. \tag{3.3}$$

From (3.2) and (3.3) we see that the probability that the individuals involved in the event are taken from the i current survivors is

$$\frac{i\nu+i(i-1)\lambda}{M[\nu+(M-1)\lambda]}. \tag{3.4}$$

Conditioning on the event whose probability is given in (3.4) we conclude from

(3.2) and (3.3) that, if stage i occurs at time τ, the probability that it involves the immigration of any particular survivor is

$$\frac{1}{i}\,\frac{\nu}{\nu+(i-1)\lambda} \tag{3.5}$$

whereas the probability that it involves any particular survivor as offspring and any other as its parent is

$$\frac{1}{i}\,\frac{\lambda}{\nu+(i-1)\lambda}. \tag{3.6}$$

If $\nu(\tau) = \theta\lambda(\tau)$ for some constant θ, the probabilities in (3.5) and (3.6) become independent of time and are given respectively by

$$\frac{1}{i}\,\frac{\theta}{\theta+i-1} \tag{3.7}$$

and

$$\frac{1}{i}\,\frac{1}{\theta+i-1}\ . \tag{3.8}$$

Fundamental to our argument above is the Markovian nature of the process (which also guarantees the independence of the different stages in the sample-forest) and our assumption of selective neutrality, which ensures that all individuals have the same fertility and mortality. This is our justification for supposing that all possible parents of an individual are equally probable. Note that (3.7) and (3.8) depend only on the ratio of the immigration and birth rates and are quite independent of the death rate. Also our discussion does not require the establishment of an equilibrium. We see that (3.7) and (3.8) are identical to (2.1) and (2.2), so the urn model of Section 2 can be applied to this time-dependent birth-death-immigration model.

The requirement that $\nu(t)=\theta\lambda(t)$ is not as artificial as it may seem. Generally, a more favorable environment leading to a higher birth rate may well attract a higher rate of immigration, or both rates may undergo the same periodic variation. More quantitatively, we may suppose that the "immigrants" are in fact mutant offspring. Denote by $\Lambda(t)$ the birth rate (per head) of any offspring, whether mutant or not. Suppose that any offspring, independent of any other, has probability $u(t)$ of being a (previously unknown) mutant. Then $\lambda(t)$, the birth rate of non-mutant offspring is given by

$$\lambda(t) = [1-u(t)]\Lambda(t)$$

and $\nu(t)$, the rate of appearance in the population of "immigrants" (that is, mutant offspring) is given by

$$\nu(t) = M(t)u(t)\Lambda(t).$$

Hence $\nu(t)=\theta\lambda(t)$ provided the probability of mutation is related to the population size by the same relation (3.1) as in Moran's model. This is the same sort of relationship implicit in diffusion models. The validity of such assumptions has been discussed by Karlin and McGregor (1964), Crow (1985) and Gillespie (1989).

Originally we took a sample of size m, but there is no reason why m may not be equal to the total population size M. This is also the case for Moran's model. Thus, for these models, any results derived apply not only to a sample, but also to the population as a whole.

4. Distributions of Genetic Properties. A number of authors have obtained results concerning genetic distributions in the context of various models. The purpose of this section is to show how the urn model provides via straightforward combinatorics a simple computational tool for deriving such results for any genetic model compatible with (2.1) and (2.2). Most of the results we discuss have been obtained elsewhere by other methods, but Theorem 4.12 is believed to be new. We shall refer to equations and other results in Donnelly and Tavaré (1986), Kelly (1977) and Watterson (1984) by the prefixes DT, K and W respectively.

We take a sample of size m, without replacement, and trace its ancestry back in time as previously discussed. (We again emphasize that in the birth-death-immigration and Moran models m may equal the total population size.) However for most of our results we now terminate the procedure after stage k+1 (that is, after $m-k$ stages) at a time when there are k survivors each of whom heads a genetically identical family consisting of a particular subset of the original sample. In Watterson's (1984) terminology these are "old" genes. Let $\lambda_1, \lambda_2, ..., \lambda_k$ be the sizes of these families. By this stage a number, l (say), of mutants have appeared each heading a genetically identical family of size $\mu_1, \mu_2, ..., \mu_l$. These are Watterson's "new" genes. Let z be the total number of old genes, so that

$$z = \Sigma\lambda_i = m - \Sigma\mu_i. \tag{4.1}$$

THEOREM 4.1 (DT 2.8). We label in some way the genes in the original sample of size m. The probability that, after stage k+1, the old gene families consist of specified genes and are of sizes $\lambda_1, \lambda_2, ..., \lambda_k$ and that there are l families of new genes (again consisting of specified individuals) of sizes $\mu_1, \mu_2, ..., \mu_l$ *in order of increasing allele age* (i.e. the mutant originator of the family of size μ_1 is younger than the mutant originator of the family of size μ_2, and so on) is

$$\frac{(m-k)!\ k!\ \theta^l}{m!\ (\theta+k)^{[m-k]}} \ \frac{\lambda_1!\lambda_2!\dots\lambda_k!\mu_1!\mu_2!\dots\mu_l!}{(\mu_1+\dots+\mu_l)\dots(\mu_1+\mu_2)\mu_1}, \tag{4.2}$$

where θ has the value appropriate to the particular model under consideration, as described in Section 3, and the rising factorial notation is defined by

$$(x)^{[n]} = x(x+1)\,(x+2)\dots(x+n-1).$$

We give a proof of this result via the urn model. For stage i, commencing with i balls in the urn, the probabilities that any particular ball is chosen as offspring followed by respectively the black ball or any particular parent ball are given by (2.1) and (2.2). Following the original sample through stages m, $m-1,\dots,k+1$ corresponds to operating the urn model until k balls remain in the urn, and the appearance of l mutants corresponds to the black ball's having been chosen on l occasions. Therefore, from (2.1) and (2.2), we see that the probability of any particular sequence leading to this outcome is

$$\frac{1}{m(m-1)\dots(k+1)} \ \frac{\theta^l}{(\theta+m-1)\,(\theta+m-2)\dots(\theta+k)}$$

$$= \frac{k!\,\theta^l}{m!\ (\theta+k)^{[m-k]}}. \tag{4.3}$$

In order to calculate the number of (equally probable) paths giving the result specified in the statement of the theorem, we first calculate the number of orders in which we can choose the offspring balls. We designate balls as "old" or "new" according as the corresponding genes are "old" or "new". We choose one ball from each of the old families (in $\lambda_1 \lambda_2 \dots \lambda_k$ ways). These will be the k balls finally left in the urn. We number the remaining balls 1, 2, ..., m–k as described below and withdraw the offspring balls in this order. We first number the remaining old balls at random (in (m–k)!/(m–z)! ways). We next choose one ball from each of the new families (in $\mu_1 \mu_2 \dots \mu_l$ ways). These will be the oldest members of each new family. We now number the new balls with the remaining m–z numbers starting with the oldest family (of size μ_l) and finishing with the youngest family (of size μ_1). The allocation of numbers here is random except that when we come to a fresh family we first give its oldest member the highest available number. This clearly guarantees the required age structure of the new families, and can be done in $(m - z)!/[(\mu_1 + \ldots + \mu_l)\ (\mu_1 + \ldots + \mu_{l-1}) \ldots (\mu_1 + \mu_2)\mu_1]$ ways.

Within any family the parent balls are successively chosen from a diminishing pool of $(\lambda-1),(\lambda-2), \ldots$ (or $(\mu-1), (\mu-2), \ldots$) potential parents, so the parent balls can be chosen in $(\lambda_1-1)!\ldots(\lambda_k-1)!(\mu_1-1)!\ldots(\mu_l-1)!$ ways. Putting all these factors together, we find that the number of ways we can obtain the desired outcome is

$$\frac{(m-k)!\,\lambda_1!\,\lambda_2!\ldots\lambda_k!\,\mu_1!\,\mu_2!\ldots\mu_l!}{(\mu_1+\ldots+\mu_l)\ldots(\mu_1+\mu_2)\,\mu_1}. \tag{4.4}$$

If we multiply (4.4) by the probability of any path (4.3) we obtain the result (4.2).

THEOREM 4.2 (W 2.9). The situation is the same as in Theorem 4.1 except that ages of the new families are now immaterial. The probability is now

$$\frac{(m-k)!\,k!\,\theta^l}{m!\,(\theta+k)^{[m-k]}}\,\lambda_1!\,\lambda_2!\ldots\lambda_k!\,(\mu_1-1)!\,(\mu_2-1)!\ldots(\mu_l-1)!. \tag{4.5}$$

The proof is the same as in Theorem 4.1 except for the numbering of the balls. Once we have chosen the k (old) balls to be left in the urn, we simply number at random all the remaining m-k (old and new) balls. This can be done in $(m - k)!$ ways and the result follows immediately.

THEOREM 4.3 (W 3.3.1). As in Theorem 4.2, except that we do not specify the particular genes belonging to any family. We simply require k old families of sizes $\lambda_1, \ldots, \lambda_k$ and l new families of sizes $\mu_1, \ldots, \mu_l$. Denote by α_j (β_j) the number of old (new) families of size j. The probability of this configuration is

$$\frac{(m-k)!\,k!\,\theta^l}{(\theta+k)^{[m-k]}}\,\frac{1}{\mu_1\mu_2\ldots\mu_l\;\alpha_1!\,\alpha_2!\ldots\alpha_m!\;\beta_1!\,\beta_2!\ldots\beta_m!}. \tag{4.6}$$

To prove this, we note that the number of ways m balls can be distributed amongst families of the given sizes is

$$m!/[\lambda_1!\,\lambda_2!\ldots\lambda_k!\mu_1!\mu_2!\ldots\mu_l!\alpha_1!\alpha_2!\ldots\alpha_m!\,\beta_1!\beta_2!\ldots\beta_m!]. \tag{4.7}$$

The $\lambda_i!$ or $\mu_i!$ factors account for the fact that the ordering within a family is immaterial and the $\alpha_j!$ or $\beta_j!$ factors account for the fact that the ordering of families of the same type containing the same number of members is immaterial. If we multiply (4.5) by (4.7) we obtain (4.6).

Similarly we can multiply (4.2) by $m!/[\lambda_1!\lambda_2!...\lambda_k!\mu_1!\mu_2!...\mu_l!\alpha_1!\alpha_2!...\alpha_m!]$ to obtain a corresponding result where the allocation of particular genes to families is unspecified but the age ordering of the new families is retained.

THEOREM 4.4 (W 3.4.1). The probability that, after stage k+1 (when k old families remain), there are α_j old families of size j (j=1, 2, ..., m) is given by

$$\Pr(\alpha_1,\alpha_2, ..., \alpha_m \mid k) = \frac{(m-k)!\,k!}{(\theta+k)^{[m-k]}} \binom{\theta+m-z-1}{m-z} \frac{1}{\alpha_1!\,\alpha_2!\ldots\alpha_m!} \qquad (4.8)$$

where $z = \Sigma j\alpha_j$ and $k = \Sigma\alpha_j$.

First, for a particular division into old and new balls, we obtain the probability of a particular sequence of offspring balls and old parent balls, summed over all possibilities for new parent balls. The probability of picking any particular sequence of m–k offspring balls is $[m(m-1)...(k+1)]^{-1} = k!/m!$. Suppose that after i stages we have removed j balls destined to be new so that the urn contains m–i balls of which $m-z-j$ are destined to be new. After the next offspring ball has been chosen, there will be $m-i-1$ balls in the urn, to which is added the black ball. A second ball is now withdrawn. If the offspring ball was destined to be old, the probability of picking any particular parent ball destined to be old is $(\theta + m - i - 1)^{-1}$. On the other hand, if the offspring ball was destined to be new, the second ball picked from the urn may be the black ball or any of the $m-z-j-1$ remaining balls that are potentially new parents. The sum of the corresponding probabilities is $(\theta+m-z-j-1)/(\theta+m-i-1)$. Bearing in mind that there are a total of $m-k$ stages of which m–z involve new balls, we see that the probability of a particular sequence of old parent balls summed over all possibilities for new parents is

$$\frac{(\theta+m-z-1)\,(\theta+m-z-2)\ldots\theta}{(\theta+m-1)\,(\theta+m-2)\ldots(\theta+k)} = \frac{(m-z)!}{(\theta+k)^{[m-k]}} \binom{\theta+m-z-1}{m-z}. \qquad (4.9)$$

We must now calculate how many such sequences give rise to an outcome conforming to the statement of the theorem. We can choose the old balls in $\binom{m}{z}$ ways. These balls can be distributed in families of the required sizes, $\lambda_1, \lambda_2, ..., \lambda_k$ in $z!/(\lambda_1!\lambda_2!...\lambda_k!\alpha_1!\alpha_2!...\alpha_m!)$ ways (compare (4.7)). The k balls which are finally left in the urn, one from each old family, can be chosen in $\lambda_1\lambda_2 ...\lambda_k$

ways. As before, the offspring balls may be picked in $(m-k)!$ different orders, whereas the old parent balls may be chosen in $(\lambda_1 - 1)!...(\lambda_k - 1)!$ different orders. Assembling all the factors and simplifying gives (4.8).

THEOREM 4.5 (W 3.6.2, but note that Watterson's equation apparently contains an error which we correct here.). The probability that after stage k+1 there are l new families of sizes $\mu_1, ..., \mu_l$ such that β_j of these families have j members (j=1,2,...,m) is given by

$$\Pr(\beta_1,\beta_2,...,\beta_m \mid k) = \frac{(m-k)!\,\theta^l}{(\theta+k)^{[m-k]}} \binom{z-1}{z-k} \frac{1}{\mu_1\mu_2\cdots\mu_l\beta_1!\,\beta_2!\cdots\beta_m!}, \tag{4.10}$$

where $l = \Sigma\beta_j$ and z is given by (4.1).

The argument parallels that of the previous theorem. First, for a particular division into old and new balls, we find the probability of a particular sequence of offspring balls and new parent balls, summed over all possibilities for old parent balls. The probability of any particular offspring sequence is, as before, k!/m!. Suppose now that after i stages we have removed j balls destined to be *old*, leaving z–j. If the next offspring ball to be chosen is destined to be new, the second ball chosen will be either black, with probability $\theta/(\theta+m-i-1)$, or any particular parent ball destined to be new, with probability $(\theta+m-i-1)^{-1}$. If the offspring ball was destined to be old, the parent ball may be any of the (z–j–1) remaining balls destined to be old. Since the black ball will be chosen l times and a total of z–k old parents will be chosen, the probability of a particular sequence of new parent balls summed over all possibilities for old parents is

$$\frac{(z-1)(z-2)\ldots k \quad \theta^l}{(\theta+m-1)(\theta+m-2)\ldots(\theta+k)} = \frac{(z-1)!}{(k-1)!}\,\frac{\theta^l}{(\theta+k)^{[m-k]}} \tag{4.11}$$

We can choose the old balls in $\binom{m}{z}$ ways, and from these the k old balls finally remaining in the urn can be chosen in $\binom{z}{k}$ ways. The m–z new balls can be distributed in families of the required sizes in $(m-z)!/(\mu_1!\mu_2!...\mu_l!\,\beta_1!\beta_2!...\beta_m!)$ ways. Finally, the offspring balls may be picked in $(m-k)!$ different orders and the new parent balls in $(\mu_1-1)!(\mu_2-1)!\,...(\mu_l-1)!$ different orders. Putting these factors together and simplifying gives (4.10).

THEOREM 4.6 (W 3.4.5) The probability that after stage k+1 there are a total of z old genes (and hence m–z new genes) is given by an inverse hypergeometric distribution:

$$\Pr(z \mid k) = \binom{z-1}{k-1}\binom{\theta+m-z-1}{m-z} \Big/ \binom{\theta+m-1}{m-k}. \tag{4.12}$$

The proof combines arguments from Theorems 4.4 and 4.5. For a particular division into old and new balls, we find the probability of a particular sequence of offspring balls, but now summed over all possibilities for parent balls. The probability of any particular offspring sequence is k!/m!. The arguments leading up to (4.9) and (4.11) show that the sum of all probabilities for the $z-k$ old parents and m–z new parents or black balls is

$$\frac{[(z-1)(z-2)\ldots k]\,[(\theta+m-z-1)(\theta+m-z-2)\ldots\theta]}{(\theta+m-1)(\theta+m-2)\ldots(\theta+k)} \tag{4.13}$$

$$= \frac{(z-1)!\,(m-z)!}{(k-1)!\,(m-k)!}\;\frac{\binom{\theta+m-z-1}{m-z}}{\binom{\theta+m-1}{m-k}}.$$

As before we can choose the old balls in $\binom{m}{z}$ ways, from these the k old balls finally remaining in the urn can be chosen in $\binom{z}{k}$ ways, and we can pick the offspring balls in $(m-k)!$ different orders. These factors together give (4.12).

THEOREM 4.7 The probability that after stage k+1 there are z old genes and l families of new genes is given by

$$\Pr(z, l \mid k) = \binom{z-1}{k-1}\frac{\theta^{l}\,c(m-z,\,l)}{(m-z)!} \Big/ \binom{\theta+m-1}{m-k}, \tag{4.14}$$

where $c(m-z, l)$ is a signless Stirling number of the first kind defined by

$$(x)^{[n]} = \sum_{l=0}^{n} c(n, l)\,x^{l}.$$

The argument is identical to that for the previous theorem except for the choice of new parent balls. Now the black ball is chosen on l occasions and a non-black ball on the remaining $(m-z-l)$ occasions. Hence the factor $(\theta+m-z-1)(\theta+m-z-2)\ldots\theta$ in (4.13) must be replaced by

$$\theta^{l} \sum a_1 a_2 \ldots a_{m-z-l}$$

where the sum is taken over all a_i satisfying

$$m - z - 1 \geq a_1 > a_2 > \ldots > a_{m-z-l} \geq 1$$

and is equal to c(m–z, l). This replacement transforms (4.12) into (4.14).

THEOREM 4.8 (W 3.4.6).

$$\Pr(\alpha_1,\alpha_2,\ldots,\alpha_m \mid z,k) = \frac{k!}{\alpha_1!\,\alpha_2!\ldots\alpha_m!} \Big/ \binom{z-1}{k-1}. \tag{4.15}$$

THEOREM 4.9 (W 3.5.1).

$$\Pr(\beta_1,\beta_2,\ldots,\beta_m \mid z, k) = \frac{(m-z)!\,\theta^{l}}{\theta^{[m-z]}} \quad \frac{1}{\mu_1\mu_2\ldots\mu_l\ \beta_1!\,\beta_2!\ldots\beta_m!}. \tag{4.16}$$

As Watterson points out, this is Ewens' (1972) formula applied to the m-z new genes.

THEOREM 4.10 (W 3.5.2).

$$\Pr(l \mid z,k) = \theta^{l} c(m-z, l)/\theta^{[m-z]}. \tag{4.17}$$

These three theorems follow immediately from, respectively, Theorems 4.4, 4.5 or 4.7 and Theorem 4.6.

Similarly, Theorems 4.3 and 4.6 show that $\Pr(\alpha_1,\alpha_2,\ldots,\alpha_m,\beta_1,\beta_2,\ldots,\beta_m \mid z,k)$ is given by the product of the right-hand sides of (4.15) and (4.16) and, using (4.17) and (4.16), we can readily obtain an expression for $\Pr(\beta_1,\beta_2,\ldots,\beta_m \mid l,z,k)$.

In any of the foregoing theorems we may of course set k equal to zero to obtain results corresponding to a complete tracing of the sample's ancestry through m stages until no old families remain.

THEOREM 4.11 (DT, Remark, page 14; K, Result 7). We take (without replacement) a sample of size m. (As we have previously emphasized, in appropriate models m may equal the total population size.) Tracing the ancestry of the sample through m stages we discover as usual a number of distinct families each of which was founded some time in the past by the introduction of a previously unknown allele. The probability that the oldest allele in the sample is represented by z members of the sample is

$$(\theta/m)\binom{m}{z} \Big/ \binom{\theta+m-1}{z}.$$

This is found, after a little manipulation, from (4.12) with k set equal to one, since, when only one old family remains, this family necessarily contains the oldest allele in the sample.

THEOREM 4.12 The probability that the youngest allele in the sample is represented by y members of the sample is given by a truncated Yule distribution:

$$\frac{\theta}{y} \Big/ \binom{\theta+y}{y} \qquad y = 1,2,\ldots, m-1 \qquad (4.18)$$

$$\binom{\theta+m-1}{m-1}^{-1} \qquad y = m. \qquad (4.19)$$

Suppose that after stage k+1 there is one new family and it has y members. This could happen either because such a family originated at stage k+1 or because there was one new family with y members at the end of stage k+2, and the parent ball at stage k+1 was non-black (with probability k/(θ+k)). Hence the probability that the youngest family in the sample contains y members and was introduced into the sample-forest at stage k+1 is

$$\Pr(z=m-y,\, l=1 \mid k) - \Pr(z=m-y,\, l=1 \mid k+1)[k/(\theta+k)].$$

Using (4.14) and the fact that c(y,1)=(y–1)! this can be written as

$$\left[\frac{y}{\theta+k}\binom{m-y-1}{k-1}\binom{\theta+y}{y}\Big/\binom{\theta+m-1}{m-k-1}\right]\left[\frac{\theta}{y}\Big/\binom{\theta+y}{y}\right]. \qquad (4.20)$$

If y=m, then necessarily k=0 and (4.20) reduces to (4.19). If y<m then to obtain the required probability we must sum (4.20) over the possible values of $k=1,2,\ldots,m-y$. The terms in the first square bracket in (4.20) are in the form of an inverse hypergeometric distribution in $k-1$ and therefore sum to one, leaving the second square bracket which is equal to (4.18).

THEOREM 4.13 (DT Proposition 4.1; K, Result 10). Take a sample of size m and trace its ancestry back through m stages, thereby partitioning the sample into genetically distinct families. Suppose one such family contains j members. Then the allele carried by this family is the oldest in the sample with probability j/m.

The urn model provides a trivial proof of this result. The oldest allele in the sample will be that carried by the last ball to be removed from the urn. Each ball has equal probability 1/m of being the last to be removed from the urn. Therefore a family with j members contains the last ball to be removed with probability j/m.

5. Relationship to Other Urn Models. Ewens (1972) laid the foundation for the use of urn models in population genetics in his treatment of the infinite-alleles Wright-Fisher model at equilibrium. His intuitive argument showed that, in drawing a sample one by one from the population, the probability that the ith gene drawn is of a novel allelic type not seen on the first $i-1$ draws is $\theta/(\theta+i-1)$, independent of the allelic composition of the first $i-1$ genes.

The urn idea was made explicit by Hoppe (1984) and Watterson (1985). The ingredient they added was the specification of the allelic type of the ith gene if it is not novel: they realized that its type is simply uniformly chosen from the genes (not alleles) already present in the sample. Specifically, the probability that the ith gene is of a particular allelic type already existing in the sample is $\mu/(\theta+i-1)$, where μ is the number of genes amongst the first $i-1$ to be drawn that are of that particular allelic type. Thus, in order to simulate a sample of size m, Hoppe considers an urn originally containing a black ball of mass θ. At each of m successive steps a ball is drawn from the urn (with probability proportional to its mass). If the black ball is drawn, it is replaced together with a ball (of unit mass) of a new colour; if a coloured (not black) ball is drawn, it is replaced together with an identical ball. After m steps the allelic composition of the sample is represented by the colour composition of the non-black balls in the urn. Watterson's construction is equivalent.

This model was investigated in considerable detail by Donnelly (1986) and Hoppe (1987). They showed that the urn model mimics the drawing of a sample, at some fixed time, from a Poisson-Dirichlet population, but they also established a connection with the time development of the underlying stochastic process: take a sample of size m and completely trace its coalescent backwards in time through m stages until no old genes remain; Donnelly and Hoppe show that Hoppe's urn model (for a sample of size m) is equivalent to the *reversed* jump chain of the coalescent. The labelling of the colours of the first m balls in his urn model according to the order of their appearance is probabilistically equivalent to labelling the allelic types in the time-reversed coalescent according to their appearance in genealogical time, i.e. according to their relative ages. This correspondence allowed both Donnelly and Hoppe to rederive in a simple fashion a number of known results.

By contrast, the urn model described in Section 2 directly mimics the coalescent itself (rather than its time reversal). If necessary, it can also keep track of the ages of alleles (as, for example, in Theorems 4.1, 4.11, 4.12 and 4.13). If the coalescent runs its full course, leaving no old genes, (that is, in our notation, k=z=0) then it is often unimportant which way we consider time to run and many results can be obtained equally well by Hoppe's model or by our model. For example results 3, 9 and 11 of Section 5 of Hoppe (1987) are respectively the same as our Theorem 4.13, Theorem 4.10 (with k=z=0, m=2, l=1) and Theorem 4.10 (with k=z=0). However, the new feature of our model is that it provides a simple computational tool in those cases (such as those considered by Watterson (1984) and Donnelly and Tavare (1986)) where, in tracing backwards through time, the coalescent does not run its full course and so there are some old genes remaining. This corresponds, in our model, to the urn being only partly emptied.

In a more recent paper, Watterson (1989) considers the situation where θ varies in time because of a varying population size or mutation rate (or both). There is nothing intrinsic to the coalescent or to our urn model that requires θ to be constant; we simply require all genes, at any given time, to be equally likely to die, give birth or mutate. Therefore we can adapt our model to include the case discussed by Watterson. Suppose θ is piecewise constant, taking values $\theta_1,\theta_2,\theta_3,...$ during successive time periods of duration $t_1,t_2,t_3,...$ proceeding from the present into the past. Let l_1, l_2, l_3,... be the (random) numbers of survivors - in the terminology of Section 2 - at the times when the value of θ changes. Watterson discusses the distribution and simulation of these numbers. Conditional on the values of l_1, l_2, l_3,... we can reproduce Watterson's result (2.2.11) by operating our urn model in the usual fashion, but changing the relative probability of picking the black ball from θ_k to θ_{k+1} after $m - l_k$ stages, where m is the size of the sample. As usual, our model can be terminated before the urn is empty thereby producing results concerning the joint distribution of old and new genes.

Watterson (1987) has discussed yet another urn model which he uses to simulate samples from a population with both neutral and deleterious allelic types. The coalescent, and hence our urn model, are based in an essential fashion on selective neutrality and hence cannot be used in this case.

References

[1] Crow, J. F. (1985). The neutrality-selection controversy in the history of evolution and population genetics. In *Population Genetics and Molecular Evolution* (T. Ohta and K. Aoki, eds.) 1-18. Japan Sci. Soc. Press, Tokyo/ Springer-Verlag, Berlin.

[2] Donnelly, P. (1986). Partition structures, Polya urns, the Ewens sampling formula, and the ages of alleles. *Theoret. Population Biol.* 30 271-288.

[3] Donnelly, P. and Tavaré, S. (1986). The ages of alleles and a coalescent. *Adv. in Appl. Probab.* 18 1-19.

[4] Ewens, W. J. (1972). The sampling theory of selectively neutral alleles. *Theoret. Population Biol.* 3 87-112.

[5] Gillespie, J. H. (1989). When not to use diffusion processes in population genetics. In *Mathematical Evolutionary Theory* (M. W. Feldman, ed.) 57-70. Princeton U.P.

[6] Hoppe, F. M. (1984). Polya-like urns and the Ewens' sampling formula. *J. Math. Biol.* 20 91-94.

[7] Hoppe, F. M. (1987). The sampling theory of neutral alleles and an urn model in population genetics. *J. Math. Biol.* 25 123-159.

[8] Karlin, S. and McGregor, J. (1964). On some stochastic models in genetics. In *Stochastic Models in Medicine and Biology* (J. Gurland, ed.) 245-279. Univ. of Wisconsin Press, Madison.

[9] Kelly, F. P. (1977). Exact results for the Moran neutral allele model. *Adv. in Appl. Probab.* 9 197-201.

[10] Kingman, J. F. C. (1982a). On the genealogy of large populations. *J. Appl. Probab.* 19A 27-43.

[11] Kingman, J. F. C. (1982b). Exchangeability and the evolution of large populations. In *Exchangeability in Probability and Statistics* (G. Koch and F. Spizzichino, eds.) 91-112. North-Holland, Amsterdam.

[12] Kingman, J. F. C. (1982c). The coalescent. *Stochastic Process. Appl.* 13 235-248.

[13] Tavaré, S. (1989). The genealogy of the birth, death, and immigration process. In *Mathematical Evolutionary Theory* (M. W. Feldman, ed.) 41-56. Princeton U.P.

[14] Watterson, G. A. (1984). Lines of descent and the coalescent. *Theoret. Population Biol.* 26 77-92.

[15] Watterson, G. A. (1985). Estimating species divergence times using multi-locus data. In *Population Genetics and Molecular Evolution* (T. Ohta and K. Aoki, eds.) 163-183. Japan Sci. Soc. Press, Tokyo/Springer-Verlag, Berlin.

[16] Watterson, G. A. (1987). Estimating the proportion of neutral mutants. *Genet. Res., Camb.* 501 155-163.

[17] Watterson, G. A. (1989). The neutral alleles model with bottlenecks. In *Mathematical Evolutionary Theory* (M. W. Feldman, ed.) 26-40. Princeton U. P.

Figure 1

A diagram showing part of a population history. Time runs down the page. At a birth, the parent is shown by a vertical line, the offspring by a slanting line. Mutant births are denoted by black circles. A sample of six individuals is denoted by black squares, and the sample-fores arising therefrom by thick lines.

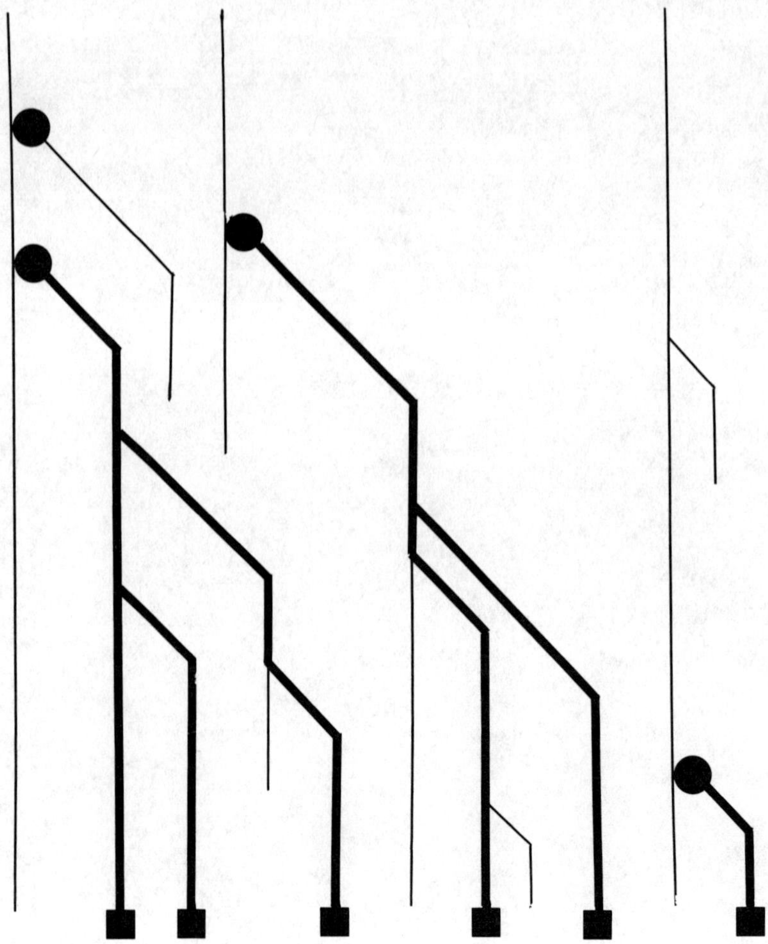

The Genealogy of Patterns of ESS's

C. Cannings and G. T. Vickers.
University of Sheffield

Summary

In a finite conflict a given payoff matrix **A** may possess **ESS** 's with supports $T_1, T_2, ..., T_k$. Such a set of supports is called a pattern and is attainable since there exists a payoff matrix **A** with **ESS** 's with those supports. This paper presents a summary of previous work aimed at specifying the set of attainable patterns, the most recent results for conflicts with 5 pure strategies and defines the notion of a genealogy of patterns. This genealogy, in which a subpattern is regarded as an 'offspring' is displayed for n = 4 and n = 5 where the genealogy has been produced so as to minimize the number of line-intersections using simulated annealing.

Introduction. A finite conflict is defined by a pair $\{\mathbf{U}, \mathbf{A}\}$ where $\mathbf{U} = \{1, 2, 3 \ldots n\}$ is the strategy space i. e. there are n pure strategies labeled 1 through n, and **A** is an $n \times n$ matrix whose elements a_{ij} are the payoffs; a_{ij} is the payoff to an individual who plays strategy i and whose opponent plays strategy j. We consider, as in classical game theory (Von Neumann and Morgenstern,1953), the mixed extension so that the strategies are $\mathbf{p} = (p_1, p_2 \ldots p_n)$ where $p \in \Delta, \Delta = \{p; p \geq 0, \Sigma p_i = 1\}$. The payoff for strategy **p** against **q** has the expected value $\mathbf{p}^T \mathbf{Aq}$ and this is also the payoff of strategy **p** in a population which is playing **q** on average.

An **ESS**, evolutionarily stable strategy, is a **p** such that for every $\mathbf{q} \neq \mathbf{p}$

(1)$\mathbf{E(p, p)} \geq \mathbf{E(q, p)}$ and

(2) if equality in (1) then $\mathbf{E(p, q)} > \mathbf{E(q, q)}$.

For the generic case of a finite conflict, a strategy **p** is an **ESS** iff (if and only if) $\mathbf{E}(i, \mathbf{p}) = \mathbf{E(p, p)}$ all $i \in \mathbf{R(p)}$, where $\mathbf{R(p)} = \{i: p_i > 0\}$ the support of **p**, $\mathbf{E}(j, \mathbf{p}) < \mathbf{E(p, p)}$, all $j \notin \mathbf{R(p)}$, and if $\mathbf{B} = (a_{ij})$, $i, j \in \mathbf{R(p)}$ then $\mathbf{z^tBz}$ is negative definite where **z** such that $\Sigma z_i = 0$, Haigh(1975).

If **P** is the power set of **U** and $\mathbf{T} = \{T_1, T_2 \ldots T_k\} \subset P$ then **T** is said to be a pattern, and in the context of conflict theory **T** is said to be an attainable pattern if there exists an **A** (a real $n \times n$ matrix) such that there are precisely k **ESS**'s for **A** and the supports are $\mathbf{T}_i$, $i = 1,2...k$ Vickers and Cannings(1988a). A pattern $\mathbf{V} = \{V_1, V_2 ... V_l\}$ is said to be a subpattern of $\mathbf{T} = \{T_1, T_2 ... T_k\}$ if $k > l$ and for some permutation of $\{1,2, ... l\}$, $\{i_1, i_2 ... i_l\}$ $\mathrm{V_j} = T_{i_j}$ $\mathrm{j} = 1, ... l$. In this case we write $\mathbf{V} \subset \mathbf{T}$. A fundamental conjecture of Vickers and Cannings(1988a) is that if **T** is attainable and $\mathbf{V} \subset \mathbf{T}$ then **V** is attainable(for $n \geq 2$). In a series of papers Cannings and Vickers(1988, 1989), Vickers and Cannings (1988a, 1988b) the authors have begun an attempt to specify the details of the set of attainable patterns. Results have been essentially of two types; exclusion results which specify certain features which an attainable pattern may not possess, and existence results which demonstrate how the payoff matrix can be constructed to achieve specific classes of patterns. These two constitute the main building blocks of the theory. However the theory is far from complete; there are many patterns which are neither excluded nor demonstrated by these results. This mortar has to be filled in by ad hoc methods and there are still many gaps in the edifice. After briefly reviewing the results previously published, and adding some new ones, we shall consider the problem of representing the whole genealogy of attainable patterns for $n = 4,5$. For given n the genealogy of attainable patterns is the digraph $\mathbf{G} = \{\mathrm{V, E}\}$ where the vertex-set **V** consists of all attainable patterns(if there is uncertainty then all patterns which have not been shown unattainable are included) and the edge-set $\mathbf{E} = \{(\mathbf{U}, \mathbf{W}) : \mathbf{U} \in \mathbf{V}, \mathbf{W} \in \mathbf{V}, \mathbf{U} \subset \mathbf{W}\}$. Essentially we regard **W** as a 'parent' of **U** and **U** as an 'offspring' of **W**. In contrast to genealogies of biological organisms, at least those known to man, an individual here may have many parents. **G** is of course just a digraph but since (**V**, descendant of) is a poset it seems natural to refer to it as a genealogy and exploit the 'generations' which, see below, exist.

2. Exclusion Results. (1) Bishop and Cannings(1976) proved that if **p** is an **ESS** with support **R(p)** and **q** has $\mathbf{R}(\mathbf{q}) \subset \mathbf{S}(\mathbf{p})$, where $\mathbf{S}(\mathbf{p}) = \{i: i \in \mathbf{U}, \mathbf{E}(i, \mathbf{p}) = \mathbf{E}(\mathbf{p}, \mathbf{p})\}$ then **q** is not an **ESS**. In the generic case $\mathbf{R}(\mathbf{p}) = \mathbf{S}(\mathbf{p})$ so we have that if **p** and **q** are **ESS**'s then neither $\mathbf{R}(\mathbf{p}) \subset \mathbf{R}(\mathbf{q})$ nor $\mathbf{R}(\mathbf{q}) \subset \mathbf{R}(\mathbf{p})$. Therefore an attainable pattern is an antichain (see Anderson(1988)).

The four following results are given in Vickers and Cannings(1988a). Here $\mathbf{Q} \subset \mathbf{U}$ and $(1, 2) \cup \mathbf{Q}$ is written as $(1, 2, \mathbf{Q})$ etc.

(2)If $\mathbf{Q} \subset \mathbf{U} \setminus (1,2,3)$, $T_1 = (1,2,\mathbf{Q})$, $T_2 = (1,3,\mathbf{Q})$, $T_3 = (2,3,\mathbf{Q})$, then no pattern $\{T_1, T_2, T_3\}$ is attainable.

Special cases of interest

(*a*) $\mathbf{Q} = \emptyset$ so for any triplet $(i, j, k) \subset \mathrm{U}$ the pairs (i,j), (i,k) and (j,k) cannot all occur as supports for **ESS**'s.

(*b*) $\mathbf{Q} = \mathrm{U}\backslash(1,2,3)$ then clearly there can be at most two supports with n - 1 elements.

(3) If $\mathrm{Q} \subset \mathrm{U}\backslash(1,2,3...l)$, $T_i = (i,\mathbf{Q})$, $i = 1,...l$ and $T_{l+1} = (1,2,3...l)$then no pattern $\mathrm{T} = (T_1,T_2...T_l,T_{l+1})$ is attainable.

(4) If $\mathrm{Q} \subset \mathrm{U}\backslash(1,2,3...l)$, $T_i = (i, l, \mathbf{Q})$, $i = 1,...l - 1$ and $T_l = (1,2,3...l - 1, \mathbf{Q})$ then no pattern $\mathrm{T} = (T_1,T_2...T_l)$ is attainable.

(5) If $\mathrm{Q} \subset \mathrm{U}\backslash(1,2,3)$, $T_1 = (1,\mathrm{Q})$, $T_2 = (2,3,\mathbf{Q})$, $T_3 = (1,2)$ and $T_4 = (1,3)$ then no pattern $\mathrm{T} = (T_1,T_2,T_3,T_4)$ is attainable. This last result is given in Cannings and Vickers(1989).

3. Maximal Patterns for n=2, 3, 4. An attainable pattern is said to be maximal if it is not a proper subpattern of any attainable pattern. Thus if the conjecture above holds we need only specify the maximal patterns. The application of the results (1) to (5) above leave 2, 4 and 9 possibilities respectively for n=2, 3, and 4. These have all been shown to be attainable and the patterns are as follows:

$n = 2$ {(1,2)}, {(1),(2)}

$n = 3$ {(1,2,3)},{(1,2),(1,3)},{(1,2),(3)},{(1),(2),(3)}

$n = 4$ {(1,2,3,4)},{(1,2,3),(1,2,4)},{(1,2,3),(2,4),(3,4)}

{(1,2,3),(4)},{(1,2),(1,3),(1,4)},{(1,2),(2,3),(3,4),(1,4)}

{(1,2),(1,3),(4)},{(1,2),(3),(4)},{(1),(2),(3),(4)}.

We shall term an attainable pattern $\mathrm{T} = (T_1,T_2,T_3...T_k)$ degenerate if for any $\mathrm{V} \subset \mathrm{U}$, $\bigcup_{i \in V} T_i$ is disjoint from $\bigcup_{i \notin V} T_i$. Thus the maximal nondegenerate patterns are

$n = 2$ {(1,2)}

$n = 3$ {(1,2,3)},{(1,2),(1,3)}

$n = 4$ {(1,2,3,4)},{(1,2,3),(1,2,4)},{(1,2,3),(2,4),(3,4)}

{(1,2),(1,3),(1,4)},{(1,2),(2,3),(3,4),(1,4)}.

Any degenerate pattern is clearly attainable if the patterns in **V** and not -**V** are attainable, and not otherwise.

4. Existence Results Suppose that **C** is the class of matrices such that for $A \in \mathbf{C}$ we have $a_{ij} = a_{ji}$, $a_{ii} = 0$, $a_{ij} = +1$ or -1. Then Cannings and Vickers(1988) prove that the supports of the **ESS**'s of any $A \in \mathbf{C}$ are precisely the cliques (maximal complete subgraphs) of the graph **G** = (**U**,**E**) where $(i,j) \in \mathbf{E}$ iff $a_{ij} = +1$. This class generates all the above examples with the exception of {(1,2,3),(2,4),(3,4)}. In addition we can easily see that any pattern with $|T_i| = 2$ all i is attainable by a matrix of **C** with **G** having an edge matching each T_i iff that G is triangle-free. Since (2) of the exclusion results asserts that there can be no triplets with all three pairs we see that that is the only restriction in this set of patterns. We note in passing that the extensive literature on graph theory can be exploited in various ways to yield information on patterns; for example random graphs can represent possible evolutions of conflicts, Cannings and Vickers(1988).

5. New Results. We present here some new results with respect to **ESS** patterns for the case n=5. These add to the previously most complete set of information which was given in Cannings and Vickers(1989).

5.1 410A *Lemma 1.* If $bf + ad < df$ or $de + bc < be$ or $fc + ae < ac$ or $a + c < 0$ or $b + e < 0$ or $d + f < 0$ then

$$\begin{vmatrix} 0 & a & b \\ c & 0 & d \\ e & f & 0 \end{vmatrix}$$

has no internal **ESS** .

Proof. The conditions are a subset of those of Theorem 3 of Vickers and Cannings(1988).

Lemma 2. The above matrix is equivalent to a symmetric matrix iff

$$a+d+e=b+c+f.$$

Lemma 3. If a,b,c,d,e,f are all positive, then the above matrix cannot have 2 ESS's.

Proof. The only possibility to be excluded is that it has two **ESS**'s of size 2, with support {1,2} and {2,3} say. However if strategy 1 cannot invade the **ESS** with support {2,3} then $f > a$ and this implies that strategy 3 can invade the **ESS** with support {1,2}.

Lemma 4. If a and c have opposite sign and if d and f have opposite sign, then the above matrix does not have an internal **ESS** if it is equivalent to a symmetric matrix.

Proof. We argue by contradiction. With the assumptions regarding the signs of a,c,d,f the matrix has one of the following sign arrangements:

$$\begin{matrix} 0 & + & \\ - & 0 & + \\ & - & 0 \end{matrix} \qquad \begin{matrix} 0 & + & \\ - & 0 & - \\ & + & 0 \end{matrix} \qquad \begin{matrix} 0 & - & \\ + & 0 & - \\ & + & 0 \end{matrix} \qquad \begin{matrix} 0 & - & \\ + & 0 & + \\ & - & 0 \end{matrix}$$

The last of these can be eliminated because it has a pure strategy 2 as an **ESS** and hence no internal **ESS**. The signs of b and e can now be inferred for the remaining three cases by applying

(a) no pure **ESS** is allowed so each column must contain at least one +,

(b) $a + d + e = b + c + f$ (from Lemma 2) The three possibilities are

$$\begin{matrix} 0 & + & + \\ - & 0 & + \\ + & - & 0 \end{matrix} \qquad \begin{matrix} 0 & + & + \\ - & 0 & - \\ + & + & 0 \end{matrix} \qquad \begin{matrix} 0 & - & + \\ + & 0 & - \\ + & + & 0 \end{matrix}$$

For the first of these we have that $de + bc > be$ and hence $d > b$ thus $a + d + e = b + c + f$ gives $0 < a + e < c + f < 0$. The third matrix is dealt with in a similar fashion. For the second we see that $de + bc > be$ implies an immediate contradiction.

Theorem. The pattern **410A** (see Vickers and Cannings,1988 for the key) is not attainable by any symmetric matrix.

Proof. Suppose the contrary. Then the matrix

$$\begin{vmatrix} 0 & \alpha & \beta & & \\ \gamma & 0 & a & b & \\ \delta & c & 0 & d & \theta \\ & e & f & 0 & \phi \\ & & \chi & \omega & \end{vmatrix}$$

can have ESS's with supports {1,2} , {1,3} , {2,3,4} , {3,5} , {4,5} . By the first lemma, a and c are not both negative and by the third they are not both positive. Hence a and c have opposite signs and likewise so do d and f. Lemma 4 thus implies the given contradiction.

The pattern **410A** can be obtained e.g. by the following matrix

$$\begin{vmatrix} 0 & 3.1 & 10 & \alpha & \alpha \\ 3.1 & 0 & 2 & -1 & \alpha \\ 4 & -1 & 0 & 2 & 3.1 \\ \alpha & 2 & -1 & 0 & 4 \\ \alpha & \alpha & 3.1 & 10 & 0 \end{vmatrix}$$

provided that α is sufficiently large and negative when the **ESS's** are specified by the columns of

$$\begin{vmatrix} 1/2 & 5/7 & 0 & 0 & 0 \\ 1/2 & 0 & 1/3 & 0 & 0 \\ 0 & 2/7 & 1/3 & 1/2 & 0 \\ 0 & 0 & 1/3 & 0 & 2/7 \\ 0 & 0 & 0 & 1/2 & 5/7 \end{vmatrix} .$$

This example and the theorem correct an error in Vickers and Cannings(1988a). In that paper the matrix which was given as corresponding to the pattern **410A** has in fact the pattern **410C** . The theorem demonstrates that symmetric matrices (which are relevant to the single-locus multi-allelic problem) will not show the full range of patterns available to general matrices.

5.2 111 The pattern **111** can be produced by the symmetric matrix

$$\begin{vmatrix} 20 & 28 & 18 & 18 & 20.5 \\ 28 & -44 & 50 & 50 & -180 \\ 18 & 50 & 0 & 16 & 50 \\ 18 & 50 & 16 & 0 & 50 \\ 20.5 & -180 & 50 & 50 & 8 \end{vmatrix}$$

and the ESS's are specified by the columns of

1/4	0	25/26
1/4	0	0
1/4	1/4	0
1/4	1/4	0
0	1/2	1/26

5.3 021A The pattern **021A** is produced by the symmetric matrix

-146.82	77.27	66.54	60.78	-61.70
77.27	-9.92	-5.56	-3.44	3.29
66.54	-5.56	-3.34	-0.25	2.41
60.78	-3.44	-0.25	0.32	1.46
-61.70	3.29	2.41	1.46	0.49

and the **ESS**'s are specified by the columns of

0.272654	0.0	0.0
0.594505	0.117229	0.0
0.093022	0.085717	0.1763
0.039819	0.0	0.156412
0.0	0.797055	0.667287

6 Genealogy for n=4. For n=4 we saw above that there were only five maximal non-degenerate attainable patterns. However there are eleven non-degenerate patterns and these are shown in Figure 1 in their genealogy. This genealogy has been drawn in generations , each generation corresponding to the number of elements in the pattern. The form shown also has the desirable property, from the point of view of drawing, that none of the edges cross. Note that the founders are the maximal patterns and that non-founders have one, two or three parents.

7 Genealogy for n=5. For n=5 the application of all our methods plus some special cases leaves us with potentially 61 patterns of which 55 are known to be attainable, there being 20 maximal ones of which 16 are known attainable. There are 6 generations and the geneaology is shown in Figure 2. Simulated annealing (see next section) allowed the production of a genealogy which had 165 intersections of the edges. In an attempt to produce a representation which is easier to read the genealogy was split into two pieces, the top four generations and the bottom three, so that the same individuals appear at the bottom of the first and the top of the second segments. This resulted in some reduction of the total number of intersections there being 102 and 35 intersections in the two segments. The reduction is less than might have been anticipated the top segment being fairly similar to the top of the whole genealogy and so is not shown here.

8 Simulated Annealing. Simulated annealing is a probabalistic optimisation technique (e.g. Lundy,1985). Thomas(1989) has applied it to the production of diagrams of genealogies. His technique involves imposing a requirement that nodes be seperated by some minimal distance and then minimising the sum of squared distances between all pairs of nodes. The criterion adopted here is rather different. Each node is assigned to its generation and within a generation nodes are numbered (i. e. assigned an ordering), the objective function is taken as the number of intersections between edges. Thus two edges (i, j) and (k, l), where i is a parent of j and k a parent of l, intersect only if i and k are in the same generation (j and l are then necessarily in the same generation), and either $m(i) < m(k)$ and $m(j) > m(l)$, or $m(i) > m(k)$ and $m(j) < m(l)$, where $m(i)$ is the position of i in its generation list.

References

[1] Anderson, I (1988). *Combinatorics of Finite Sets.* Oxford University Press.

[2] Bishop, D. and Cannings, C. (1976). Models of animal conflicts. *Adv.Appl.Prob.*, **8**, 616-621.

[3] Cannings, C. and Vickers, G. T. (1988). Patterns of **ESS** 's II. *J. Theoret. Biol.*, **132**, 409-420.

[4] Cannings, C. and Vickers, G. T. (1989). Patterns and invasions of evoutionarily stable strategies. *J. Appl. Maths and Computation.*,**32**,227-253.

[5] Haigh, J. (1975). Game theory and evolution. *Adv.Appl.Prob.*, **7** ,8-11.

[6] Lundy, M(1985). Applications of the annealing algorithm to combinatorial problems in statistics. *Biometrika*, **72** ,191-198.

[7] Thomas, A. (1988). Drawing pedigrees. I.M.A.J.Maths.Appl.Med. and Biol., **5** ,201-213.

[8] Vickers, G. T. and Cannings, C. (1988a). Patterns of **ESS** 's I. *J. Theoret. Biol.*, **132**, 387-408.

[9] Vickers, G. T. and Cannings, C. (1988b). On the number of stable equilibrium in a one-locus , multi-allelic system., **131** ,273-277.

[10] Von Neumann, J. and Morgenstern, O.(1953). *Theory of Games and Economic Behaviour* , Princeton University Press.

Figure 1

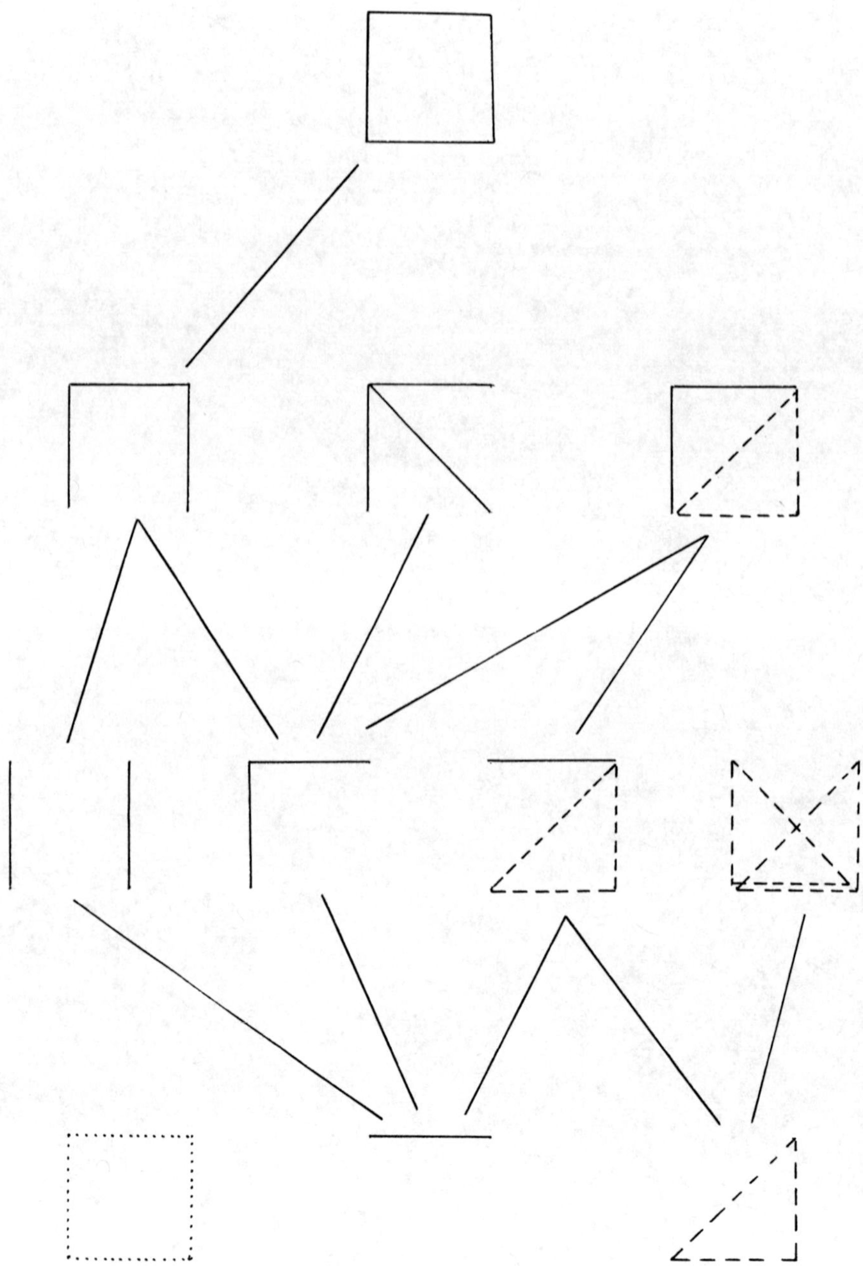

Figure 2, Part 1

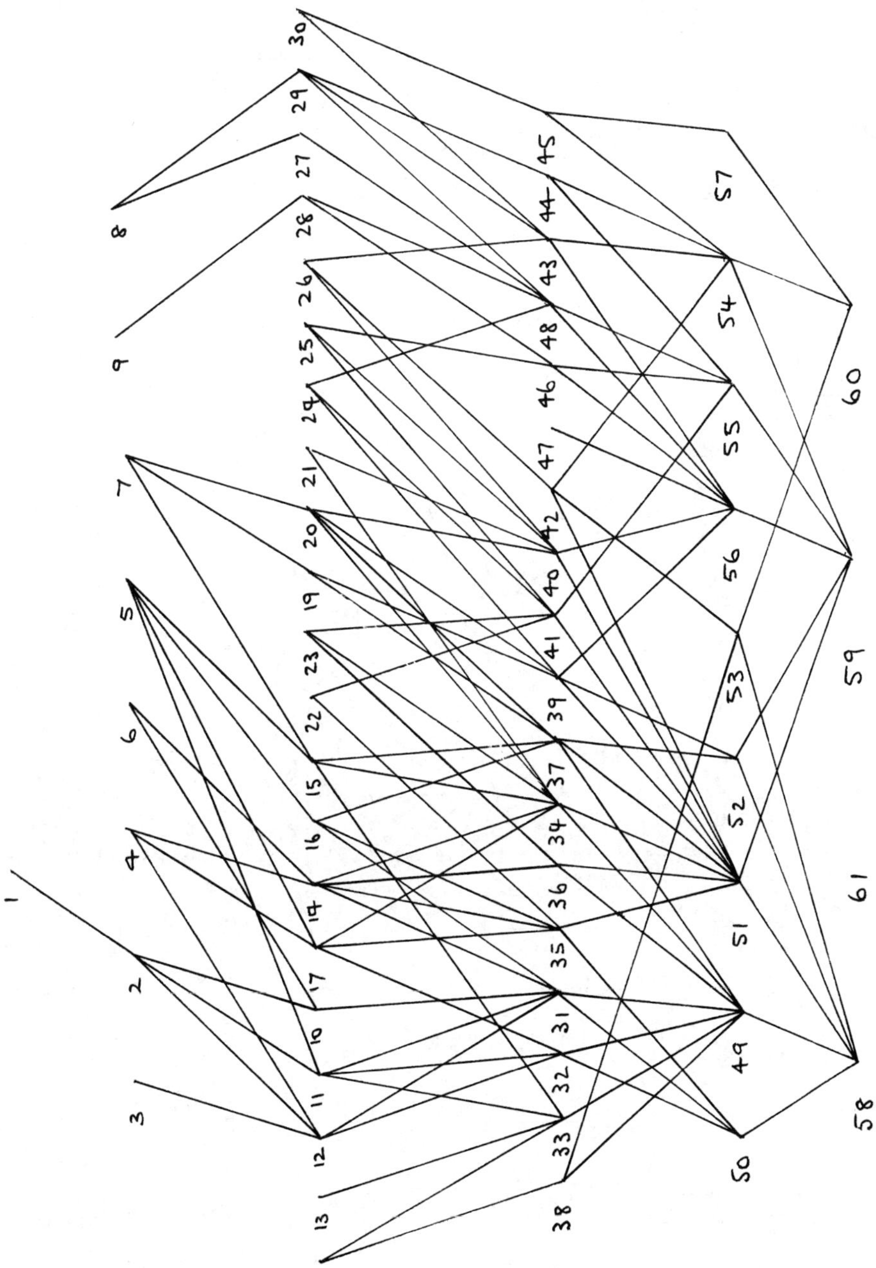

Figure 2, Part 2

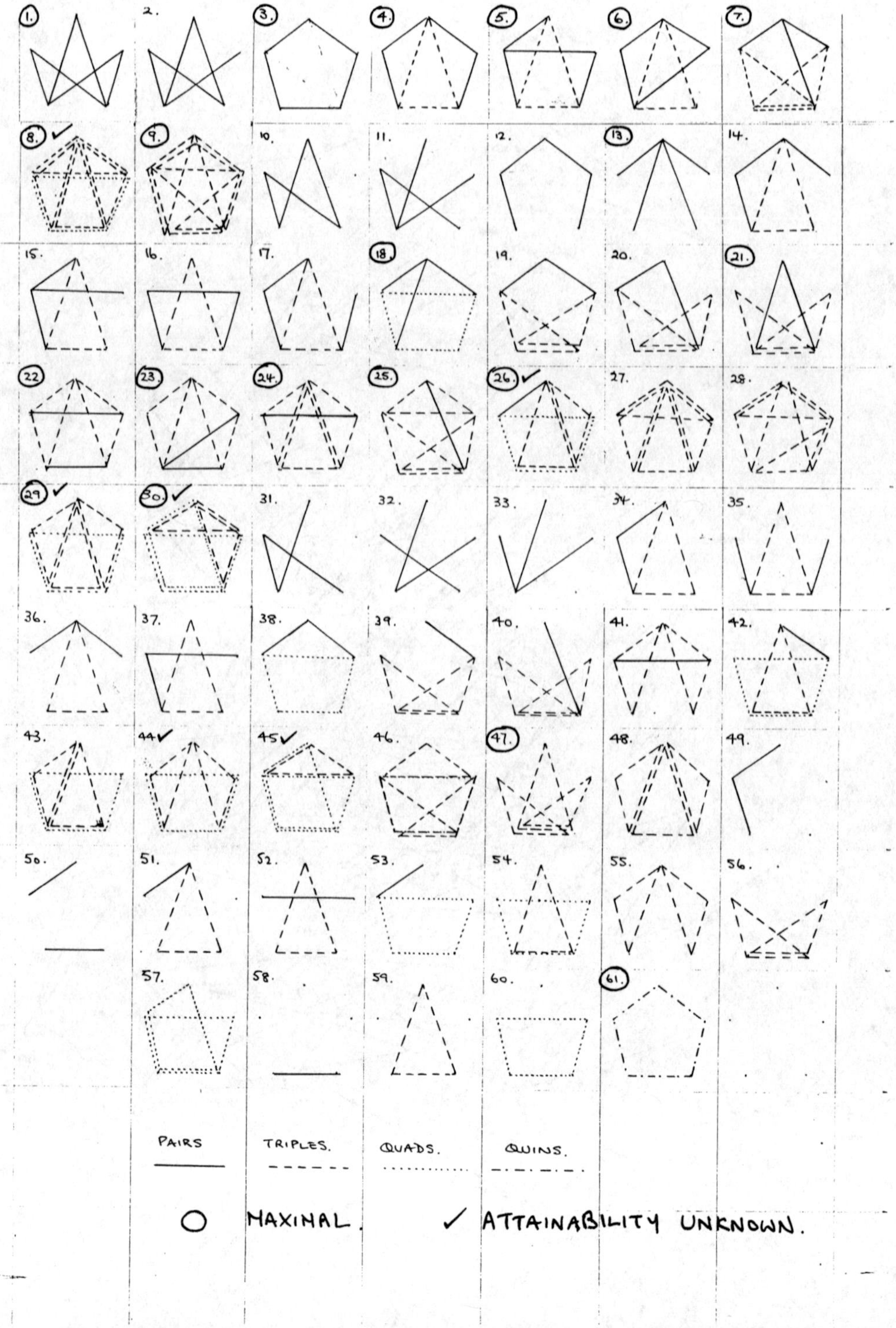

EXTINCTION PROBABILITIES OF BRANCHING PROCESSES IN RANDOM ENVIRONMENTS

D.R. Grey and Lu Zhunwei
University of Sheffield

Abstract

In the supercritical branching process with independent and identically distributed environments, it is shown that under certain regularity conditions there exists a parameter $\theta_0 > 0$ such that the probability of extinction starting with k individuals, q_k, is asymptotically of order not less than $k^{-\theta_0}$ and of smaller order than $k^{-\theta}$ for any $\theta < \theta_0$. An application to the optimal choice of strategy for minimizing the probability of extinction is mentioned.

1. Introduction and Statement of Results. We consider a branching process $\{Z_n; n = 0,1,2...\}$, where Z_n denotes the population size at time n. Reproduction is affected by a sequence of environment variables $\bar{\zeta} = \{\zeta_0, \zeta_1, \zeta_2, ...\}$ in the following way: for each n, conditional on $\bar{\zeta}$ and $Z_0, Z_1, ... Z_n$, the family sizes of the Z_n individuals at time n are independent random variables each with a distribution which is determined by ζ_n, and whose probability generating function (p.g.f.) we shall denote by ϕ_{ζ_n}. Then Z_{n+1} is just the sum of these family sizes.

Particular models for the environment variables are a sequence of independent identically distributed (i.i.d.) random variables (Smith and Wilkinson [4]) and, more generally, a stationary ergodic sequence (Athreya and Karlin [1]). In this paper, we consider only the Smith-Wilkinson model, although subsequent work has generalized the results to certain types of Athreya-Karlin model.

Let $q(\bar{\zeta})$ be the probability, conditional on $\bar{\zeta}$, that the population becomes extinct starting with a single ancestor:

$$q(\bar{\zeta}) = P(Z_n \to 0 \text{ as } n \to \infty | \bar{\zeta}, Z_0 = 1).$$

Then, because, conditional on the environment sequence, lines of descent are independent, the unconditional probability of extinction starting with k ancestors is

$$q_k = E[q(\bar{\zeta})^k]$$

where the expectation is taken over the environment.

We shall only be interested in the *supercritical* case where $P(q(\bar{\zeta}) < 1) > 0$, or equivalently $q_k < 1$ for all $k \geq 1$; well-known sufficient conditions for supercriticality are

(i) $0 < E[\log\phi'_{\zeta}(1)] < \infty$

and (ii) $E[-\log(1-\phi_{\zeta}(0))] < \infty$

and we shall adopt these conditions. Note that since the ζ_n are i.i.d. we may drop the suffix n in the above expressions.

Write $\xi = \log\phi'_{\zeta}(1)$, the log mean family size for environment ζ, and let $F(\theta) = E(e^{-\theta\xi})$, the Laplace transform of the distribution of ξ. By condition (i) above we have $F'(0) < 0$, and since F is convex there may exist $\theta_0 > 0$ such that $F(\theta_0) = 1$ and $F'(\theta_0) < \infty$; if θ_0 exists, it is unique. Under this condition, we have the following result.

Theorem. *Let $\theta_0 > 0$ exist such that $F(\theta_0) = 1$ and $F'(\theta_0) < \infty$ where F is as defined above. Then*

(a) $\liminf_{k\to\infty} k^{\theta_0} q_k > 0$;

(b) if in addition $E[(1-\phi_{\zeta}(0))^{-\theta_0}] < \infty$, then

$$q_k = o(k^{-\theta}) \text{ as } k \to \infty \text{ for all } \theta < \theta_0.$$

This gives a fairly precise idea of the asymptotic behaviour of q_k as $k \to \infty$. Note that in the Galton-Watson (constant environment) case, $\{q_k\}$ is a geometric sequence, which is possible since θ_0 does not exist, as $F(\theta) = e^{-\theta\xi}$ for some constant ξ. However, if the environment may be either favourable ($\xi > 0$) or unfavourable ($\xi < 0$) then the existence of θ_0 is not unreasonable as a regularity condition.

In Section 2 we shall prove the above theorem and in Section 3 we shall describe an application to an optimal strategy for survival.

2. Proof of the Theorem. Intuitively, if the population size is large then, by the law of large numbers, its fluctuations are almost entirely determined by the environment variables, and, more specifically, $\{\log Z_n\}$ behaves rather like a random walk with increments $\log\phi'_{\zeta_n}(1) = \xi_n$. Therefore, the probability of extinction should behave like the probability that a random walk with positive drift ever reaches a low level, and this is the key to our proof.

However, rather than work with $\{\log Z_n\}$ it is easier to consider the *dual process* $\{X_n\}$ defined by Smith and Wilkinson [4]:

$$X_{n+1} = \phi_{\zeta_n}(X_n)$$

with $X_0 = 0$. This is a Markov process with state space $[0,1[$ and with a proper limiting/equilibrium distribution in the supercritical case. Moreover, if X is a random variable with this distribution then we may write

$$q_k = EX^k$$

since in fact X has the same distribution as $q(\bar{\zeta})$. Thus, the behaviour of q_k as $k \to \infty$ may be investigated via the behaviour of the distribution of X close to 1.

We make the transformations

$$Y_n = -\log(1 - X_n)$$

$$\text{and } Y = -\log(1 - X).$$

Clearly $\{Y_n\}$ is also a Markov process and Y possesses its limiting/equilibrium distribution. We are interested in the tail behaviour of this distribution at $+\infty$.

$$\begin{aligned} Y_{n+1} &= -\log(1 - \phi_{\zeta_n}(X_n)) \\ &= -\log(1 - X_n) \quad + \left(-\log\left(\frac{1 - \phi_{\zeta_n}(X_n)}{1 - X_n}\right)\right) \\ &\geq Y_n + (-\log\phi'_{\zeta_n}(1)) \qquad \text{since } \phi_{\zeta_n} \text{ is a p.g.f.} \\ &= Y_n + (-\xi_n). \end{aligned}$$

Also $Y_{n+1} \geq 0$. Hence

$$Y_{n+1} \geq \max(0, Y_n + (-\xi_n)).$$

Define $W_0 = 0$ and

$$W_{n+1} = \max(0, W_n + (-\xi_n)) \qquad \text{for } n = 0,1,2,\ldots$$

Then $\{W_n\}$ is a random walk with left reflecting barrier at 0 and jumps $\{-\xi_n\}$. It is easy to show by induction on n that $Y_n \geq W_n$ for all n, and thence that

$$P(Y \geq y) \geq P(W \geq y)$$

for all $y \geq 0$, where W is a random variable with the limiting/equilibrium distribution of $\{W_n\}$.

The following duality argument is familiar:

$$W_n = \max(0, -\xi_{n-1}, -\xi_{n-1} - \xi_{n-2}, \ldots, -\xi_{n-1} - \ldots - \xi_0)$$

$$= -\min(0, \xi_0, \xi_0 + \xi_1, \ldots, \xi_0 + \ldots + \xi_{n-1}) \text{ in distribution}$$

so that $-W$ has the same distribution as the all-time minimum M of an unrestricted random walk with jumps $\{\xi_n\}$. But by the assumption of the existence of θ_0 and the result of Feller [2], Ch. XII, we know

$$P(M \leq -t) \sim ce^{-\theta_0 t}$$

as $t \to \infty$, for some constant $c > 0$. Hence

$$P(Y \geq y) \geq P(W \geq y) \geq c' e^{-\theta_0 y}$$

for all $y \geq 0$ for some constant $c' > 0$. In terms of X, this becomes

$$P(X \geq x) \geq c'(1-x)^{\theta_0}$$

for $0 \leq x < 1$. Hence, denoting the distribution function of X by H,

$$q_k = EX^k = \int_0^1 x^k dH(x)$$

$$= k\int_0^1 x^{k-1}(1 - H(x))\,dx$$

$$\geq c'k\int_0^1 x^{k-1}(1-x)^{\theta_0} dx$$

$$= c'kB(k, \theta_0 + 1)$$

where B is the Beta function. Finally, since

$$B(k, \theta_0 + 1) \sim \frac{\Gamma(\theta_0 + 1)}{k^{\theta_0 + 1}}$$

as $k \to \infty$, we deduce that $q_k \geq c'' k^{-\theta_0}$ for all k, some $c'' > 0$, and part (a) of the theorem is proved.

To prove part (b) of the theorem, fix $x_0 \in [0, 1[$ and let $y_0 = -\log(1 - x_0)$.

Also let

$$\hat{\xi}_n = \log\left(\frac{1-\phi_{\zeta_n}(x_0)}{1-x_0}\right)$$

and define $U_0 = y_0$ and

$$U_{n+1} = \max(y_0, U_n + (-\hat{\xi}_n))$$

so that $\{U_n\}$ is a random walk with left reflecting barrier at y_0 and jumps $\{-\hat{\xi}_n\}$. We show by induction on n that $Y_n \le U_n$ for all n. Obviously $Y_0 = 0 \le y_0 = U_0$. If $Y_n \le U_n$ then either $Y_n \le y_0$ in which case

$$\begin{aligned} Y_{n+1} &= -\log(1-\phi_{\zeta_n}(X_n)) \\ &\le -\log(1-\phi_{\zeta_n}(x_0)) \qquad \text{since } X_n \le x_0 \\ &= y_0 + \left(-\log\left(\frac{1-\phi_{\zeta_n}(x_0)}{1-x_0}\right)\right) \\ &\le U_n + (-\hat{\xi}_n) \\ &\le U_{n+1} \end{aligned}$$

or, alternatively, $y_0 \le Y_n \le U_n$ in which case

$$\begin{aligned} Y_{n+1} &= Y_n + \left(-\log\left(\frac{1-\phi_{\zeta_n}(X_n)}{1-X_n}\right)\right) \\ &\le Y_n + \left(-\log\left(\frac{1-\phi_{\zeta_n}(x_0)}{1-x_0}\right)\right) \qquad \text{since } \phi_{\zeta_n} \text{ is a p.g.f.} \\ &\le U_n + (-\hat{\xi}_n) \\ &\le U_{n+1}. \end{aligned}$$

It follows using similar arguments to those used in the proof of part (a), but with the inequalities reversed, that if $\theta > 0$ can be found such that $\{U_n\}$ has an equilibrium distribution whose tail is of order $e^{-\theta y}$ as $y \to \infty$, then $q_k = O(k^{-\theta})$ as $k \to \infty$ for any such θ. We show that for all choices of x_0 sufficiently close to 1, such $\theta = \theta(x_0)$ can indeed be found, and moreover that $\theta(x_0) \uparrow \theta_0$ as $x_0 \uparrow 1$, which is sufficient to establish part (b) of the theorem.

Let

$$F_x(\theta) = E\left(e^{-\theta\xi}\right) = E\left(\frac{1-\phi_\zeta(x)}{1-x}\right)^{-\theta}$$

where, for ease of notation, we replace x_0 by x and suppress the suffix n. By the properties of p.g.f.'s,

$$\phi'_\zeta(1)^{-\theta} \le \left(\frac{1-\phi_\zeta(x)}{1-x}\right)^{-\theta} \le (1-\phi_\zeta(0))^{-\theta}$$

and so, using the extra condition of the statement of part (b) of the theorem, we know that $F_x(\theta)$ is finite for $0 \le \theta \le \theta_0$. Also, using dominated convergence, $F_x(\theta) \quad E(\phi'_\zeta(1))^{-\theta} = F(\theta)$ as $x \quad 1$ for each $\theta \in [0, \theta_0]$. Hence, since $F(\theta) < 0$ for $0 < \theta < \theta_0$, it follows that for all x sufficiently close to 1, $F_x(\theta) < 0$ for some θ and therefore, since F_x is convex, $F'_x(0) < 0$; moreover, $F_x(\theta_0) \ge 1$ and so there exists $\theta = \theta(x) > 0$ such that $F_x(\theta) = 1$ and obviously $F'_x(\theta) < \infty$. The fact that $\theta(x) \uparrow \theta_0$ as $x \uparrow 1$ also follows easily from the fact that $F_x(\theta) \downarrow F(\theta)$ as $x \uparrow 1$. This completes the proof of the theorem.

3. Application to an optimal strategy for survival. In Grey [3] a model was described in which laying birds may choose randomized strategies for clutch size and thereby effectively choose between different family size distributions, in an attempt to minimize the probability of extinction of the species. It was stated loosely that maximizing the parameter θ_0 which appears in this paper should be optimal for large populations, and an intuitive justification was given. A numerical example showed that this criterion may lead to a genuinely randomized strategy, which may be interpreted as hedging one's bets against unknown future environments.

The results of this paper provide a sound theoretical justification for the criterion suggested above, insofar as they prove that between any two strategies with different values of θ_0, the one with the larger value will indeed yield a smaller probability of extinction provided that the initial population size is sufficiently large.

References

[1] Athreya, K.B. and Karlin, S. (1971). On branching processes in random environments I & II . *Ann. Math. Stat.* **42** , 1499-1520 & 1843-1858.

[2] Feller, W. (1971). *An Introduction to Probability Theory and its Applications* Vol. II, Second Edition. Wiley, New York.

[3] Grey, D.R. (1980). Minimisation of extinction probabilities in reproducing populations. *Theoretical Population Biology* **18** , 430-443.

[4] Smith, W.L. and Wilkinson, W.E. (1969). On branching processes in random environments. *Ann. Math. Stat.* **40** , 814-827.

MEASURE-VALUED PROCESSES: TECHNIQUES AND APPLICATIONS

Kenneth J. Hochberg[1]
Bar-Ilan University

Abstract

Several techniques are presented for the analysis of measure-valued stochastic processes. These methods are then applied to a number of examples so as to determine the behavior of the processes at fixed times, in the long term, and in the renormalization limit.

I. Introduction. Interest in the theory of measure-valued stochastic processes, which were first introduced by Dawson (1975) in the study of branching diffusion systems, has recently increased as a result of the work of Dynkin (1988, 1989) and Perkins (1988, 1989, 1990) on superprocesses. In the interim, measure-valued processes have been used to describe the dynamics of populations whose underlying distributions are continuously changing, and which are therefore described via a distribution or random measure at each fixed time. They also arise as the diffusion approximation to certain real-valued processes describing spatially-distributed systems. Applications that lead to measure-valued processes in the diffusion limit include models that describe the behavior of systems of branching and diffusing particles (Dawson (1977), Dawson-Hochberg (1979), Hochberg (1980, 1983), Iscoe (1986, 1988), Dawson-Iscoe-Perkins (1989)); models describing frequency distributions of alleles in neutral, non-neutral and interactive populations (Fleming-Viot (1979), Dawson-Hochberg (1982, 1983), Hochberg (1986), Ethier-Kurtz (1987), Ethier-Griffiths (1987, 1990), Vaillancourt (1990a,b)); and the continuous limit of hierarchically-structured branching and branching diffusion systems (Dawson-Hochberg-Wu (1990), Dawson-Hochberg (1991)).

In what follows, we review several techniques for analyzing measure-valued stochastic processes, present examples in which such processes arise, and show how these techniques are used to study the behavior of such processes at fixed times, in the long term, and in the renormalization limit.

1. Research partially supported by U.S. National Security Agency grant MDA 904-88-H-2044 and U. S. National Science Foundation grant DMS - 8800289.

II. Techniques. As is the case for any general Markov process, a measure-valued process can be described via an infinitesimal generator G, defined so as to incorporate the evolutionary forces that lead to changes in state, and a domain $D(G)$ on which the generator is defined. In the diffusion approximation, one considers an appropriately rescaled limiting version G of a generator G_N, defined for fixed values of N. One shows the existence and uniqueness of the resulting process as a Markov process with values in an appropriate space of measures via semigroup techniques or by solving an initial-value martingale problem.

For a locally compact complete separable metric space S, let $M(S)$ denote the space of bounded Radon measures on S, furnished with an appropriate metric so as to be a complete, separable metric space with a topology equivalent to that of weak convergence of measures. For a closed subset E of $M(S)$, let $B(E)$ denote the σ-algebra of Borel subsets of E, $C_b(E)$ denote the space of continuous bounded functions on E, and $L_\infty(E)$ denote the space of bounded measurable functions on E. Let $\Omega_c^E = C([0,\infty),E)$ be the space of continuous functions mapping $[0,\infty)$ into E, and let $\Omega_D^E = D([0,\infty),E)$ be the space of functions mapping $[0,\infty)$ into E that are right continuous with limits from the left. A *measure-valued* (or *E-valued*) stochastic process $\{X(t)\colon t \geq 0\}$ is then given by a mapping $X\colon\ [0,\infty]\times\Omega_C^E(\text{or } \Omega_D^E) \to E$ defined by the canonical process $X(t,\omega) \equiv \omega(t)$ for $t \geq 0$, $\omega \in \Omega_C^E(\text{or } \Omega_D^E)$. The distribution of a measure-valued stochastic process $\{X(t)\colon t \geq 0\}$ is determined by a measurable mapping $\mu \to P_\mu$ from E into $P(\Omega_C^E)$ or $P(\Omega_D^E)$, the space of probability measures on Ω_C^E and Ω_D^E, respectively, furnished with the topology of weak convergence.

An E-valued stochastic process $\{X(t)\colon t \geq 0\}$ with time-homogeneous transition probabilities is uniquely determined by the characteristic functional of the initial distribution $X(0) = \nu \in E$, defined for $f \in C_b(S)$ by

$$\phi(f) \equiv \int_E \exp\left(i\int_S f(x)\,\nu(dx)\right)P(d\nu), \tag{2.1}$$

and the characteristic functional of the probability transition function, given by

$$\phi_{t,\nu}(f) \equiv E\left\{ exp\left(i\int_S f(x)X(t,dx)\right)\middle|\, X(0) = \nu\right\}. \tag{2.2}$$

Equivalently, one can use Laplace functionals of the form

$$L_{t,\nu}(f) = E\{exp\left(-\int_S f(x)X(t,dx)\right)\Big| X(0) = \nu\}. \tag{2.3}$$

Under these circumstances, the *cumulant generating function* $u(\cdot,\cdot)$ will satisfy

$$L_{t,\nu}(f) = exp\left(-\int_S u(t,x)\nu(dx)\right), \tag{2.4}$$

where $u(t,x)$ can be obtained as the solution to a (generally nonlinear) initial-value problem of the form

$$\frac{\partial u(t,x)}{\partial t} = Au(t,x) + h(u) \tag{2.5}$$

$$u(0,x) = f(x).$$

In the semigroup approach to proving existence, the goal is now to show that the solution $u(t,x)$ can be expressed as

$$u(t,x) = U_t f(x) \quad , \tag{2.6}$$

where $\{U_t: t \geq 0\}$ is a semigroup of nonlinear operators on $C_b(S)$; it is then related to the generator G of the Markov process via the relationship $Gf = \lim_{t\, 0}(U_t f - f)/t$.

Another approach to proving existence, due to Stroock and Varadhan, is to solve an appropriate martingale problem. Specifically, for a subset A of $C_b(E) \times L_\infty(E)$, a family of probability distributions $\{P_\mu: \mu \in E\}$ is a $P(\Omega_C^E)$ – (*or* $P(\Omega_D^E)$ –) *solution to the initial-value martingale problem* for A if for every pair $(f,g) \in A$, the following hold:

(i) $P_\mu \in P(\Omega_C^E)$ (or $P(\Omega_D^E)$) for each $\mu \in E$;

(ii) $P_\mu\{\omega : X(0,\omega) = \mu\} = 1$ for each $\mu \in E$;

(iii) for each P_μ, the canonical process $\{X(t): t \geq 0\}$ is a *solution to the martingale problem* for A in the sense that for every $(f,g) \in A$,

$$Z(t) \equiv f(X(t)) - \int_0^t g(X(s))\,ds$$

is a P_μ-martingale.

Dawson and Kurtz (1982) proved that the uniqueness, measurability, and strong Markov property of a $P\{\Omega_C^{M(S^*)}\}$-valued solution $\{P_\mu: \mu \in M(S^*)\}$ of the initial-value martingale problem associated with the pair (F_f, GF_f) for $f \in D = \bigcup_{N=0}^{\infty} D(S^{*N})$, where S^* denotes the one-point compactification $S \cup \{\infty\}$ of S, $F_f(\mu)$ is the *monomial*

$$F_f(\mu) \equiv \int_S \ldots \int_S f(x_1, \ldots x_{N(f)})\, \mu(dx_1) \ldots \mu(dx_{N(f)}) \tag{2.7}$$

on $M(S^*)$, $N(f) = N$ if $f \in D(S^{*N})$, and S^{*N} denotes the N-fold Cartesian product of S, follows from the existence of a solution to the martingale problem associated with the function-valued dual process $\{Y(t): t \geq 0\}$ with generator G^d defined via the relationship

$$GF_f(\mu) = G^d F_\mu(f) + V(N(f))\, F_\mu(f) \tag{2.8}$$

for a function V defined on the set $\{0,1,2,\ldots\}$, where $F_f(u) \equiv F_\mu(f)$. $X(t)$ and $Y(t)$ are said to form a pair of *dual processes* if the martingale problem associated with $\{(F_f, G F_f) : f \in D(G)\}$ has an $M(S^*)$-valued solution $\{X(t): t \geq 0\}$ and the martingale problem associated with $\{(F_\mu, G^d F_\mu): \mu \in E\}$ has a $\Pi^D(E)$-valued solution $\{Y(t): t \geq 0\}$, where $\Pi^D(E)$ denotes the smallest algebra of functions on E which contains $\Pi_1^D(E)$, the set of monomials with coefficients $f \in D$ restricted to E.

The function-valued process $Y(t)$ has a state space $\Pi^D(E)$ that is sufficiently rich so as to generate $L_\infty(E)$ under bounded pointwise convergence, so one can apply the theorem of Stroock and Varadhan (1979) that asserts that there will be at most one solution to the initial-value martingale problem for A if for each $t \geq 0$ and $F \in \Pi^D(E)$, there is a function $H_{F,t} \in L_\infty(E)$ such that for $\mu \in E$,

$$E_\mu\{F_f(X(t))\} = E\{F_\mu(Y(t))\, exp\left(\int_0^t V(Y(u))\, du\right) \Big| \, Y(0) = f\}. \tag{2.9}$$

for any solution $\{P_\mu: \mu \in E\}$ of the martingale problem for A. From the duality relationship (2.8) above, it follows that the function $H_{F,t}(\mu)$ will be given by

$$E_{\mu}\{F_f(X(t))\} = E\{F_{\mu}(Y(t)) \exp\left(\int_0^t V(Y(u))\,du\right) \Big| Y(0) = f\}. \tag{2.10}$$

III. Examples. A. *Measure-valued branching diffusion process.* Branching diffusion processes consist of individuals or particles that move in space according to some deterministic diffusion, and independently, at random times, each particle either splits into k ($k = 2,3,...$) particles with probability $p_k \geq 0$ or disappears with probability $p_0 = 1 - \sum_{k=2}^{\infty} p_k \geq 0$. The process is called *critical* if the expected number of particles remains constant, as occurs, for example, in a binary branching situation with $p_0 = p_2 = 1/2$.

Now consider a critical branching diffusion process where the particles move in R^d according to a symmetric stable diffusion of index α, where $0 < \alpha \leq 2$ (e.g. , $\alpha = 2$ corresponds to Brownian motion diffusion), and, independently, after exponentially distributed holding times, either branch or die. The "high density" measure-valued limit of this process has been called the *stochastic measure diffusion process* and is obtained by considering a succession of such branching diffusions with progressively increasing numbers of particles of successively smaller individual mass in such a way that the expected total mass remains constant.

The basic construction of this process, together with some of its fundamental properties, appears in Dawson (1975,1977); the local structure of the resulting random measures is presented in Dawson and Hochberg (1979). Here the semigroup approach is used, and it is shown via the Trotter product formula that the semigroup $\{U_t: t \geq 0\}$ of nonlinear operators defined in (2.6) can be well approximated by the alternation of the Markov semigroup $\{S_t: t \geq 0\}$ of contraction operators on $C_k(R^d)$ associated with the symmetric stable process on R^d of index α with the semigroup $\{T_t: t \geq 0\}$ of operators that determines the branching mechanism, defined for $f \in C_k(R^d)$ by

$$T_t f(x) = f(x)/[1 - i\gamma t f(x)], \tag{3.1}$$

successively over small intervals of time. Thus, the value of the process at time t is approximated by the result of successively creating particles according to the branching mechanism defined by (3.1) and then smearing them out via the symmetric stable diffusion, each over small time intervals of length t/m.

The cumulant generating function (2.6) here satisfies the nonlinear equation

$$\frac{\partial u(t,x)}{\partial t} = G_\alpha u(t,x) - \gamma u^2(t,x)\,, \tag{3.2}$$

where G_α is the infinitesimal generator of the Markov semigroup $\{S_t: t > 0\}$. From the scaling property

$$S_t f(0) = S_{rt} f_{\alpha,r}(0) \tag{3.3}$$

for symmetric stable diffusions, where $f_{\alpha,r}(u) \equiv f(r^{-1/\alpha}u)$ and $r > 0$, it follows that over a time interval of length t, particles will tend to diffuse only within a region whose diameter is of the order $t^{1/\alpha}$.

B. *Multilevel branching process*. Measure-valued processes arising in the study of dynamic multilevel information structures have been introduced by Dawson and Hochberg (1990, 1991). At each level, individual information units undergo a Galton-Watson-type branching process in which they can be copied or removed. In addition, collections of information units at a given level comprise information units at the next higher level which also, independently, undergo Galton-Watson branching. Such multilevel systems arise in the description of replication of digitized data banks and in the evolution of animal and plant populations, in which individuals evolve both at the local level and at the colony level, in the sense that genetic structures of individuals may be altered through births and deaths within a given colony, while the colony itself may disappear at random times due to natural calamities, or it may replicate, say to establish a copy of itself at a new habitat. These mutlilevel processes also arise in the description of new mutations in mitochondrial DNA, because sampling processes are taking place at both the organelle and the individual levels.

In the two-level critical binary branching case, we start with a finite system of branching particles on R^d, which can be represented as a point in $M(R^d)$. These particles then replicate to create superparticles or die, and each particle then continues to evolve independently and undergo further branching, thus yielding a random number of copies. The two-level branching process $\{X_t: t \geq 0\}$ can then be viewed as an $N(Z^+)$-valued pure jump Markov process, where $N(Z^+)$ denotes the set of integer-valued measures on Z^+. A measure in $N(Z^+)$ can be represented by a vector $\{(n_i): i \in Z^+\}$, where n_i denotes the number of superparticles consisting of exactly i particles. The four possible transitions of state in the process involve the birth or death in a superparticle of size i or the copy or disappearance of a superparticle of size i. The generator G of this

$N(Z^+)$-valued Markov processes and its domain $D(G)$ are given by

$$D(G) = \{F\colon F(\mu) \equiv f(\Sigma\varphi_i n_i) = f(\int \varphi d\mu) \equiv f(<\varphi,\mu>)\} \text{ where } \mu = \Sigma n_i \cdot \delta_i,$$

$$GF(\mu) = \gamma_2(1 + c_2/2)\int [f(<\varphi,\mu> + \varphi(x)) - f(<\varphi,\mu>)]\,\mu(dx) \tag{3.4}$$

$$+ \gamma_2(1 - c_2/2)\int [f(<\varphi,\mu> - \varphi(x)) - f(<\varphi,\mu>)]\mu(dx)$$

$$+ \gamma_1(1 + c_1/2)\int x[f(<\varphi,\mu> - \varphi(x) + \varphi(x+1)) - f(<\phi,\mu>)]\mu(dx)$$

$$+ \gamma_1(1 - c_1/2)\int x[f(<\varphi,\mu> - \varphi(x) + \varphi(x-1)) - f(<\phi,\mu>)]\mu(dx)$$

where γ_1 and γ_2 are constants denoting the particle and superparticle branching rates, and the constants c_1 and c_2 are determined by the mean offspring sizes of the level-1 and level-2 branching processes, so that $c_i = 0$ (> 0, < 0) corresponds to critical (supercritical, subcritical) behavior at level i. An analysis of the set of moment measures $\{m_n\colon n = 1,2,...\}$ given by

$$m_n(t,\mu\colon A_1, ..., A_n) \equiv E\{\prod_{i=1}^{n} X(t,A_i) \mid X(0) = \mu\} \tag{3.5}$$

yields the characterization of the process $\{X_t\}$ as the unique solution of the $N(Z^+)$-valued martingale problem associated with G.

The continuous diffusion limit of this multilevel system is given by the $M(R^d)$-valued process $\{Y_t\colon t > 0\}$ obtained as the weak limit of the process

$$Y_t^\varepsilon(A) = \varepsilon X_t^\varepsilon(A/\varepsilon), \tag{3.6}$$

where $\{X_t^\varepsilon\colon t \geq 0\}$ is the two-level process with $\gamma_1^\varepsilon = \gamma_1/\varepsilon$, $\gamma_2^\varepsilon = \gamma_2/\varepsilon$, $c_1^\varepsilon = \varepsilon c_1$, $c_2^\varepsilon = \varepsilon c_2$. The process $\{Y_t\}$ is characterized as the unique solution to the martingale problem associated with the generator G_c given by

$$G_cF(\mu) = f'(<\phi,\mu>) < L\phi,\mu> + \gamma_2 f'(<\phi,\mu>) < c_2\phi,\mu> \tag{3.7}$$

$$+ \frac{1}{2}\gamma_2 f''(<\phi,\mu>) <\phi^2,\mu>$$

where

$$L\phi = \frac{1}{2}\gamma_1 x\phi'' + \gamma_1 c_1 x\phi'. \tag{3.8}$$

Here, the cumulant generating function u(. , .) in the Laplace functional (2.4) that determines the transition probabilities on $M(R^+)$ satisfies

$$\frac{\partial u}{\partial t} = \frac{1}{2}\gamma_1 x\frac{\partial^2 u}{\partial x^2} + c_1\gamma_1 x\frac{\partial u}{\partial x} - \frac{1}{2}\gamma_2 u^2 + c_2\gamma_2 u \tag{3.9}$$

$$u(0,x) = \phi(x).$$

A non-explosion criterion to ensure existence and uniqueness for a general class of multilevel birth-and-death processes is given in Dawson, Hochberg and Wu (1990).

C. *Multilevel branching diffusion process.* We now consider a finite system of branching random walks on R^d and again assume that the entire system performs critical binary branching after exponentially-distributed holding times. Each replica then develops independently according to the branching random walk. This addition of a spatial structure to the lowest level yields a multilevel branching random walk, which can be represented as a random atomic measure $Y_a(t)$ on $M(R^d)$. This process and its continuous diffusion limit $Y(t)$ are $M_2(R^d) = M(M(R^d))$-valued processes. $Y(t)$ is characterized via the martingale problem for the limiting generator G given by

$$GF(\nu) = \langle\langle \mathcal{L}F'(\nu,\cdot), \nu\rangle\rangle + \frac{1}{2}\int\int F''(\nu;\mu_1,\mu_2)\,\delta_{\mu_1}(d\mu_2)\,\nu(d\mu_1) \tag{3.10}$$

where

$$\langle\langle H,\nu\rangle\rangle = \int H(\mu)\,\nu(d\mu) \tag{3.11}$$

for $H \in C(M(R^d))$, and the test functions $F(\nu)$ on $M_2(R^d)$ have the form

$$F(\nu) = f(\langle\langle h_1(\langle h_2,\cdot\rangle), \nu\rangle\rangle) \tag{3.12}$$

where $h_1, f \in C_b^2(R)$, $h_2 \in C^2(R^d)$, $\nu \in M_2(R^d)$, $\langle h,\mu\rangle = \int h\,d\mu$,

$$F'(\nu,\mu) = \frac{\delta F(\nu)}{\delta\nu(\mu)} = \frac{d}{d\varepsilon}[F(\nu+\varepsilon\delta_\mu)]_{\varepsilon=0} = f'(\langle\langle h_1(\langle h_2,\rangle),\nu\rangle\rangle)h_1(\langle h_2,\mu\rangle), \tag{3.13}$$

and $\mathcal{L}$ denotes the generator of the $M(R^d)$-valued branching process, so

$$\mathcal{L}F'(\nu,\mu) = f'(\langle\langle h_1(\langle h_2,.\rangle),\nu\rangle\rangle)h'_1(\langle h_2,\mu\rangle)\langle\Delta h_2,\mu\rangle \tag{3.14}$$

$$+\frac{1}{2}f'(\langle\langle h_1(\langle h_2,.\rangle),\nu\rangle\rangle)h''_1(\langle h_2,\mu\rangle)\langle h_2^2,\mu\rangle.$$

The Laplace functional of the M_2-valued process $Y(.,.)$ is given by

$$L_{t,\nu}(H) = E[exp(- \int_{M(R^d)} Y(t, d\mu)) \mid Y(0) = \nu] = exp(-\int U(t,\mu)\nu(d\mu)) \quad (3.15)$$

for $H \in C_+ \;(M(R^d))$, where

$$\frac{\partial u}{\partial t} = \mathcal{L}U - U^2. \quad (3.16)$$

D. *Fleming-Viot process and other genetical models.* Fleming and Viot (1979) introduced a model to describe the distribution of multi-dimensional genetic characteristics in large natural populations. Here, each coordinate represents a different observable genotypic or phenotypic characteristic of the individuals in the population. This model is analyzed in Dawson and Hochberg (1982, 1983), where it is shown that it arises as the weak limit of the continuous-time Ohta-Kimura ladder or stepwise-mutation model for selectively neutral allelic populations evolving under random genetic drift, via multinomial sampling from the empirical distribution of allelic frequencies in the host population, and a symmetric mutation structure, where the large population limit is taken so that the mutation rate remains constant while the incremental effect of each mutation is assumed to decrease.

A general characterization of the Fleming-Viot process that includes the possibility of selective advantage of some allelic types over others is provided in Ethier and Kurtz (1987). Let the compact metric space S denote the set of types, let the mutation rates be described by the generator A of a Feller semigroup on the space $C(S)$ of continuous functions on S, and let selection intensities be specified by a symmetric, bounded, Borel function σ on $S \times S$. The single-locus measure-valued diffusion $\{X_t: t \geq 0\}$ is the Markov process in $P(S)$, the space of Borel probability measures on S, associated with the generator G with domain $D(G)$ given by

$$D(G) = \{\phi : \phi(\mu) \equiv \prod_{i=1}^{n} \langle f_i,\mu\rangle, f_1,\dots f_n \in D(A), n \geq 1\},$$

$$G\phi(\mu) = \sum_{i<j} [\langle f_i f_j,\mu\rangle - \langle f_i,\mu\rangle \langle f_j,\mu\rangle] \prod_{l \neq i,j} \langle f_l,\mu\rangle \quad (3.17)$$

$$+\sum_i \langle Af_i,\mu\rangle \prod_{j\neq i} \langle f_j,\mu\rangle + \sum_i [\langle \sigma f_i,\mu^2\rangle - \langle f_i,\mu\rangle \langle \sigma,\mu^2\rangle] \prod_{j \neq i} \langle f_j,\mu\rangle,$$

where $\langle f,\mu\rangle = \int f d\mu$. Specifically, A can be taken to be the infinitesimal generator G_α of a symmetric stable diffusion of index α, or, for the case $\alpha = 2$, the Laplace operator Δ.

Vaillancourt (1990) has extended our study of this process to include a class of weakly-interacting Fleming-Viot processes. Specifically, he considers the case where offspring choose their genotype or phenotype from the empirical frequency distribution of types in some neighboring population, rather than from the parent or host population.

Ethier and Kurtz (1987) have analyzed the infinitely-many-alleles model of Kimura and Crow using a similar measure-valued approach necessitated by the introduction of selection into the model. Ethier and Griffiths (1987) provide the corresponding measure-valued analysis for the infinitely-many-sites model of Kimura. These last two models take into consideration the possibility that mutations do not always lead to previously existing states. Such dynamic models arising from molecular population genetics theory have gained more credence among geneticists as a result of the recent recognition of the gene as a sequence of nucleotides and the resulting advances in nucleotide sequencing.

E. *Multilocus Fleming-Viot process with recombination.* The multilocus measure-valued Fleming-Viot model incorporates the possibility of recombination between genes at different loci. In the two-locus diffusion, the state space is taken to be $R^{d_1} \times R^{d_2}$ and (3.17) is replaced by

$$
\begin{aligned}
G\prod_{i=1}^{n} &\langle f_i \times g_i, \mu\rangle \\
&= \sum_{i<j} [\langle f_i f_j \times g_i g_j,\mu\rangle - \langle f_i \times g_i,\mu\rangle \langle f_j \times g_j,\mu\rangle] \prod_{l \neq i,j} \langle f_l \times g_l,\mu\rangle \qquad (3.18) \\
&+ (1/2)\, D_1 \sum_{i=1}^{n} \langle \Delta_{d_1} f_i \times g_i,\mu\rangle \prod_{l \neq i} \langle f_l \times g_l,\mu\rangle \\
&+ (1/2)\, D_2 \sum_{i=1}^{n} \langle f_i \times \Delta_{d_2} g_i,\mu\rangle \prod_{l \neq i} \langle f_l \times g_l,\mu\rangle \\
&+ (1/2)\, \rho \sum_{i=1}^{n} [\langle f_i \times 1, \mu\rangle \langle 1 \times g_i, \mu\rangle - \langle f_i \times g_i, \mu\rangle] \prod_{l \neq i} \langle f_l \times g_l, \mu\rangle .
\end{aligned}
$$

where Δ_d denotes the d-dimensional Laplace operator, D_1 and D_2 are mutation rates, and ρ, $0 \leq \rho \leq \infty$, is a recombination rate such that $\rho = 0$ corresponds to complete linkage and $\rho = \infty$ corresponds to zero linkage.

Ethier and Griffiths (1990) have discussed the general two-locus theory in the case where the two mutation semigroups (one for each locus) are ergodic. Uniqueness of the measure-valued two-locus problem follows using the function-valued dual approach, since the recombination term introduces a linear birth rate while the random genetic drift (final) term introduces a quadratic death rate. The special case of complete linkage can also be considered as a particular example of (3.17).

IV. Results. Analysis of the measure-valued processes introduced in the last section has produced results of three distinct types: fixed-time behavior of the process, including analysis of the topological support set of the associated random measures, clustering phenomena for individuals or mass, coherent translation or dispersive behavior of mass over the state space, and distribution of the numbers and mass of surviving particles; long-term behavior of the process, including ergodic limits, stationarity of the limiting distribution, rates of growth or decay, and the wandering or coherent nature of the processes as time increases; and renormalization theorems, in which an appropriate scaling factor is incorporated into both the time and space variables, and the resulting rescaled process is shown to converge to a limit as the scaling factor increases.

A. *Fixed-time behavior.* At each fixed time t, the value of a measure-valued process is given by a random measure $X_t(\omega)$. This measure is said to be *singular* if there exists a random support for the process that has Lebesgue measure zero. In general, a set is said to be *singular* if it has zero Lebesgue measure.

The *Hausdorff-Besicovitch dimension of support* of a Borel set E is defined by

$$H\text{-}\dim E = \sup\{\beta > 0 : \lim_{\delta\downarrow 0} \inf_{E} \sum_i [d(E_i)]^\beta = \infty\} \tag{4.1}$$

where $d(E_i)$ is the diameter of the set E_i and

$$E = \{\{E_i\} : E \subset \bigcup_i E_i, d(E_i) < \delta \text{ for each } i\},$$

the set of all coverings of the set E by sets of diameter less than δ.

A fundamental relationship between the Hausdorff-Besicovitch dimension of support and singularity of a set lies in the following result:

THEOREM 4.1. *If E is a subset of R^d and the Hausdorff-Besicovitch dimension of E is less than d, then the set E is singular.*

A set B in R^d is called a *generalized random Cantor set* if B can be expressed as the intersection of an infinite decreasing sequence of sets $\{B_n: n = 0,1,2,...\}$, such that B_0 is a unit cube in R^d, and each B_n is the union of some number, say Λ_n, of disjoint subcubes of volume $(\Gamma_n)^{-d}$, where $\{\Gamma_n\}$ is an increasing sequence of non-negative integers. In other words, a generalized random Cantor set is a limit of sets B_n which, as n increases, themselves are the unions of progressively increasing numbers of subsets, each of which is successively smaller in volume. A generalized random Cantor set differs from the Cantor ternary set in that the sequence of numbers $\{\Lambda_n\}$ and $\{\Gamma_n\}$ are not necessarily deterministic but may be random, as are the locations of the surviving intervals or subcubes in the successive B_n's.

The following relationship between generalized random Cantor sets and Hausdorff dimensions is proven in Dawson and Hochberg (1982, Lemma 3.1) :

THEOREM 4.2. *If B is a generalized random Cantor set, then*

$H\text{-dim}\, B \leq \liminf_{n\to\infty} \log \Lambda_n / \log \Gamma_n.$

For the measure-valued branching diffusion, it is shown in Dawson and Hochberg (1979) that the approximating sequence of smeared particle processes described in section IIIA above can be viewed as a hierarchy of smeared clusters at different scales. The n-th scale is obtained by dividing the unit cube $V \subset R^d$ into $\Gamma_n{}^d$ equal subcubes of volume $\Gamma_n{}^{-d}$. For $m_n = [2\gamma t\, \Gamma_n{}^{\alpha}]$, where $[x]$ denotes the greatest integer less than or equal to x, one then obtains a generalized random Cantor set for which the random measure $X(t/m_n)$ consists of a Poisson number of clusters, each with total mass that is exponentially distributed with mean $\gamma t/m_n$. It then follows from Theorem 4.2 that the Hausdorff-Besicovitch dimension of the topological support of the random measure $X_t(\omega)$ is bounded above by the index α of the stable diffusion in dimension $d \geq \alpha$. Theorem 4.1 then implies that the random measure $X_t(\omega)$ is singular if $d > \alpha$.

Singularity of $X_t(\omega)$ for $d = \alpha$ ($\alpha = 1,2$) follows from the self-similarity property

$$X(t,A) \overset{d}{=} X(k^{\alpha}t, A_k)/k^{\alpha}, \tag{4.2}$$

where $x \in A_k$ if and only if $x/k \in A$. Taking A compact and $k = t^{-1/\alpha}$, this says that

$$X(t,A) \stackrel{d}{=} X(1,A_{t^{-1/\alpha}})/(1/t). \tag{4.3}$$

This expression may be thought of as describing a telescopic effect; it relates the relative density of the measure over a shrinking set at a fixed time to the measure over a fixed compact set as t increases. Since $X(t,A)$ converges to zero in probability for large t in one and two dimensions (Dawson (1977)), $X(1,\cdot)$ can have no absolutely continuous component.

For the Fleming-Viot process on R^d, let $S = R^d \cup \{\infty\}$, the one-point compactification of R^d. At fixed times t, the process is described by a random probability measure $X = X_t(\omega)$ on S. The *k-th moment measure* $M_k\,(dx_1,...,dx_k)$ is a probability measure on S^k defined by

$$E\{\prod_{j=1}^{k} <\phi_j, X>\} = \int_S ... \int_S \left(\prod_{i=1}^{k} \varphi_i(x_i)\right) M_k(dx_1,...,dx_k) \tag{4.4}$$

for $\phi_i \in C(S)$. It is proven in Dawson and Hochberg (1982, section 6) that a random probability measure X can be related via the system of moment measures $\{M_k(dx_1,...,dx_k) : k = 1,2,...\}$ to an infinite system of interacting particles via the following technique, which we shall call the *de Finetti representation of a random probability measure.*

The collection $\{M_k\}$ forms a consistent family of probability measures; moreover, the measure M_k is an exchangeable probability law on S^k, i.e., for $A_1,..., A_k \in B(S)$,

$$M_k(A_1,...,A_k) = M_k(A_{\pi(1)},...,A_{\pi(k)}) \tag{4.5}$$

for every permutation π. It follows from the Kolmogorov extension theorem that there exists a probability measure P^* on (S^∞, F^*), where F^* denotes the P^*-completion of the product σ-algebra, such that

$$P^*(A_1 \times ... \times A_k) = M_k(A_1,...,A_k); \tag{4.6}$$

moreover, P^* is the probability law of a sequence $\{Z_k: k = 1,2,...\}$ of exchangeable S-valued random variables, which can be viewed as the locations of a countable collection of particles in S. For ω in S^∞ and A in a countable algebra that generates $B(S)$, the de Finetti theorem for exchangeable random variables implies the existence of

$$Y^*(A,\omega) = \lim_{N\to\infty}\frac{1}{N}\sum_{j=1}^{N} I_A(Z_j(\omega)), \tag{4.7}$$

where $I_A(x)$ equals one if x is an element of the set A and zero otherwise. Y^* is a regular conditional probability given the σ-algebra F^*, so

$$P^*\{Z_k \in A_k, k = 1,2 \ldots \mid F^*\} = \prod_{k=1}^{\infty} Y^*(\omega, A_k) \quad P^* - \text{a.s.} \tag{4.8}$$

Moreover, there exists an extension of Y^* to a random measure Y such that Y is a version of X; i.e., for any set $B \in B(P(S))$,

$$P^*\{Y \in B\} = P\{X \in B\}. \tag{4.9}$$

For the Fleming-Viot process, the random motion of this related infinite particle system $\{Z_n\}$ is described by the motion of its k-particle subsystems, as follows: each particle performs an independent symmetric diffusion of index α on R^d, and, at a constant rate, one particle disappears and another splits into two particles, each of which continues to move independently according to the same symmetric diffusion law. If we look at the infinite system at time t_0 and trace back the genealogy of these particles to a time $t_0 - T_n^\infty$ at which the infinitely-many particles had exactly n common ancestors and now look only at these n "surviving family trees," we find that they form n random clusters. Moreover, within each cluster the radius is bounded, with the bound determined from the scaling property (3.2). As n increases, we obtain increasing numbers of such clusters, each with progressively smaller radii, so the clusters form a generalized random Cantor set. We then proceed as detailed earlier and conclude that the Fleming-Viot random measure has Hausdorff dimension bounded above by $min(d,\alpha)$ and is singular in dimensions greater than α, indicating a highly clustered distribution and, thus, a high correlation between the genetic traits.

The fact that $d = \alpha$ is in fact the exact lower bound for the Hausdorff dimension of support of these random measures X_t in dimensions $d \geq \alpha$ at fixed times is a consequence of the following result of Zähle (1988):

THEOREM 4.3. *Let X be a random measure with second moment measure $K(x,dy)$ and assume that for $E(X)$-a.e. $x \in R^d$,*

$$\int_{S(x,\varepsilon)} |x-y|^{-D} K(x,dy) < \infty$$

where $S(x,\varepsilon)$ is some sphere of radius $\varepsilon > 0$ centered at x. Then,

$$P\{X(B) > 0 \setminus \{\dim(\Phi \cap B) \geq D\}\} = 0,$$

where Φ denotes the closed support of X.

Since the second moment measure $K(x,dy)$ for both the measure-valued branching diffusion and Fleming-Viot processes satisfies

$$K(x,y) \sim |x-y|^{-(d-\alpha)} \tag{4.10}$$

for small $|x-y|$, we have

$$\int \frac{1}{|x-y|^D} \frac{1}{|x-y|^{d-\alpha}} dx \approx \int \frac{r^{d-1}}{r^{D+d-\alpha}} dr = \int_{0^+} \frac{1}{r^{D-\alpha+1}} dr \tag{4.11}$$

which converges if and only if $D < \alpha$, so by Theorem 4.3, $\dim \Phi \geq D$ for all $D < \alpha$, so $\dim \Phi \geq \alpha$.

The same technique applied to the multilevel branching diffusion system in four or more dimensions yields a lower bound of 2α for the Hausdorff dimension of the support of the random measure Y_t.

In each of these processes, there are two forces acting in opposing ways on the distribution: branching (and random genetic drift in the Fleming-Viot process) leads on occasion to disappearance of individuals from the population, whereas diffusion (and mutation in the Fleming-Viot case) leads to a spread of the process to new members. The mean distance between survivors in a branching process is of the order $t^{1/d}$ (and $t^{2/d}$ in the case of two-level branching), whereas a stable diffusion of index α spreads at the rate $t^{1/\alpha}$, so we see changes of behavior at the critical dimension $d = \alpha$ for the single-level cases and $d = 2\alpha$ for the two-level situation.

The topological support properties for the two-locus Fleming-Viot model with recombination given by (3.18) have not yet appeared in the literature. What is known to date can be summarized in the following theorem, obtained with S.N. Ethier:

THEOREM 4.4. *The topological support of the random measure $X_t^\rho(\omega)$ associated with the two-locus Fleming-Viot process with combination rate ρ characterized by the generator* (3.18) on $R \times R$ and $R^2 \times R^2$ *has the properties summarized in the following chart, where H-*dim *A is the Hausdorff-Besicovitch dimension of the set A and "singular" and "absolutely continuous" denote a.e.-properties of the random measure:*

space	complete linkage $\rho = 0$	partial linkage $0 < \rho < \infty$	zero linkage $\rho = \infty$
$R \times R$	H-dim $X_t^0(\omega) = 2$	H-dim $X_t^\rho(\omega) = 2$	H-dim $X_t^\infty(\omega) = 2$
	singular		absolutely continuous
$R^2 \times R^2$	H-dim $X_t^0(\omega) = 2$	$2 \leq H$-dim $X_t^\rho(\omega) \leq 4$	H-dim $X_t^\infty(\omega) = 4$
	singular	singular	singular

Proof: In the case of complete linkage there is no recombination, so we have one Fleming-Viot process in $2d$ dimensions ($d = 1,2$). Applying the result that H-dim $X_t(\omega) = min(d,2)$ for the Fleming-Viot process with Brownian diffusion parameter $\alpha = 2$, we get $H\text{-}dim\ X_t^0(\omega) = min(2d,2) = 2$ for $d = 1,2$. For $d = 1$ we have a two-dimensional Fleming-Viot random measure, which was shown above to be singular. For $d = 2$ the random measure $X_t^0(\omega)$ is singular by Theorem 4.1, since we are in a $2 \times 2 = 4$-dimensional situation, and the Hausdorff dimension of the support set of $X_t^0(\omega)$ is only two, which is less than this dimension.

In the case of zero linkage there is complete recombination, so the Hausdorff dimension of support is $min(d,2) \times min(d,2) = min(4,2d)$. For $d = 1$ we get H-dim $X_t^\infty(\omega) = 2$, and the random measure is absolutely continuous with respect to Lebesgue measure, as is each one-dimensional Fleming-Viot measure. For $d = 2$, the Hausdorff dimension of support is four, and the measure is singular because it is formed from two singular two-dimensional Fleming-Viot measures. It is then clear that for $0 < \rho < \infty$, the random measure $X_t^\rho(\omega)$ will have Hausdorff dimension of support equal to two in the case $d = 1$ as it is for the extremal cases $\rho = 0$ and $\rho = \infty$, and dimension bounded below by two and above by four in the case $d = 2$. Similarly, the measure $X_t^\rho(\omega)$ will be singular in the case $d = 2$. □

REMARK: It seems reasonable to conjecture that the exact Hausdorff dimension of $X_t^\rho(\omega)$ in the case of partial linkage will be a function of the recombination rate ρ.

The Fleming-Viot process satisfies an additional property as well, that of long-term compact coherence. A probability-measure-valued process $\{X(t): t > 0\}$ is said to be *coherent* if for every $\varepsilon > 0$ there exists t_0, $0 \leq t_0 < \infty$, with the property that for each $t \geq t_0$ there is a random sphere $S_\varepsilon(t,\omega)$ centered at

$$x(t) = \int_S x X(t,dx), \tag{4.12}$$

with radius

$$R_\varepsilon(t) = \int_S |x| X(t,dx) < \infty \tag{4.13}$$

that is a stationary stochastic process, such that

$$P\{X(t,\omega,S_\varepsilon(t,\omega)) \geq 1-\varepsilon\} = 1. \tag{4.14}$$

The *wandering motion* of a coherent distribution can be described by the process $\{x(t): t \geq 0\}$. The process $\{X(t)\}$ is said to be *compactly coherent* if (4.14) also holds for $\varepsilon = 0$ with some centering, not necessarily that prescribed by (4.12); thus, the sphere $S_0(t,\omega)$ contains all the mass of the process $X(t, \cdot)$ with probability one. A process that is not coherent is said to be *dispersive.*

The results of this section, excluding those already included in Theorem 4.4, can be summarized as follows:

THEOREM 4.5. *Let $X_t^{BD}(\omega)$ and $X_t^{FV}(\omega)$ denote the measure-valued branching diffusion and Fleming-Viot random measures, respectively , on R^d with spatial diffusion governed by a symmetric stable process of index α, $0 < \alpha \leq 2$, and let $X_t^{MLB}(\omega)$ and $X_t^{MBD}(\omega)$ denote the two-level branching and two-level branching diffusion random measures, respectively, at a fixed time t. Assume that each of these random measures has compact support at time $t = 0$ with probability one. Then the following hold:*

(i) *In any spatial dimension, each of these random measures has compact support, with probability one, for each fixed* $t > 0$.

(ii) *The Hausdorff-Besicovitch dimension of the topological supports of $X_t^{BD}(\omega)$ and $X_t^{FV}(\omega)$ both equal min(d,α).*

(iii) *The random measures $X_t^{BD}(\omega)$ and $X_t^{FV}(\omega)$ are singular in R^d for $d \geq \alpha$, with probability one.*

(iv) *The Fleming-Viot random measure $X_t^{FV}(\omega)$ is compactly coherent.*

(v) *The Hausdorff-Besicovitch dimension of the topological supports of $X_t^{MLB}(\omega)$ and $X_t^{MBD}(\omega)$ are bounded below by 2α in dimensions $d \geq 2\alpha$.*

B. *Long-term behavior.* We have already noted that the measure-valued branching diffusion process $X(t,A)$ on a compact set A converges to zero in probability for large t in the case of recurrent diffusion (e.g., in one and two dimensions for $\alpha = 2$). When the diffusion is transient, it is shown in Dawson (1977) that there exists a limiting steady-state random measure.

In Dawson and Hochberg (1982,1983), the single-locus Fleming-Viot process $\{X(t): t \geq 0\}$ on $S = R^d \cup \{\infty\}$ with $A = \Delta_d$, the d-dimensional Laplace operator, is related to a function-valued dual process $\{Y(t): t \geq 0\}$ with infinitesimal generator G^d corresponding to that of a Markov process on $C = \bigcup_{N=0}^{\infty} C(S^N)$, evolving according to the following mechanisms:

(i) $Y(t)$ jumps from $C(S^N)$ to $C(S^{N-1})$ for $N = 2,3...$

(ii) at the time of a jump from $C(S^N)$ to $C(S^{N-1})$, a pair $\{j,k\}$ is picked at random from $\{1,2,...,N\}$, and f is replaced by a function $\Phi_{jk}f$ of the N-1 variables $x_1, x_2, ..., x_{k-1}, x_{k+1}, ..., x_N$, the k - th variable x_k having merged with the j-th-variable x_j;

(iii) between jumps, $Y(t)$ evolves deterministically on $C(S^N)$ according to the Brownian motion semigroup H_t^N on $(R^d)^N$ given by

$$H_t^N f(x) = (2\pi t)^{-(Nd)/2} \int e^{-|x-y|^2/(2t)} f(y)\, dy; \tag{4.15}$$

(iv) once $Y(t)$ is a continuous function of only one variable, no further jumps occur.

The *empirical centered moments* of $\{X(t): t \geq 0\}$ are defined by

$$R_{k_1,...,k_d}(t) \equiv \int_{R^d} \prod_{i=1}^{d} (x_i - x_i(t))^{k_i} X(t, dx) \tag{4.16}$$

where the *empirical mean process* $x(t)$ is given by

$$x(t) \equiv (x_1(t), ..., x_d(t)) \tag{4.17}$$

where

$$x_i(t) = \int_{R^d} x_i X(t, dx), \quad i = 1, 2, ..., d. \tag{4.18}$$

If we apply the duality relationship (2.10) to $E_\mu\{ R_{k_1,\ldots,k_d}(t) \}$ and note that the number $N(f)$ of variables in the dual process $Y(t)$ decreases according to a pure death process and thus converges to one in finite time with probability one, we obtain the following:

THEOREM 4.6. *For the Fleming-Viot process* $\{X(t): t \geq 0\}$, *if*

$\int |x|^{N_0} \mu(dx) < \infty$ for $N_0 = \sum_{l=1}^{d} k_l$, *then*

(a) $E_\mu\{R_{k_1,\ldots,k_d}(t)\} < \infty$ *for* $t \geq 0$,

and

(b) $\lim_{t\to\infty} E_\mu\{R_{k_1,\ldots,k_d}(t)\} \equiv r_{k_1,\ldots,k_d}$

exists and is finite.

The random cluster

$$X^*(t, dx) = X(t, \{a - x(t) : a \in dx\}) \tag{4.19}$$

centered at the empirical mean $x(t)$ is a stationary process, and the Birkhoff ergodic theorem implies that

$$I(f) = \lim_{T\to\infty} \frac{1}{T} \int_0^T \int_{R^d} f(x)\, X^*(t, dx)dt \tag{4.20}$$

exists a.s. as a linear functional on $C(S)$ and is independent of the initial measure μ. We therefore have the following result:

THEOREM 4.7. *The centered Fleming-Viot random cluster* $X^*(t,dx)$ *satisfies*

$$\lim_{t\to\infty} E_\mu \{\int f(x)\, X^*(t, dx)\} = I(f) \tag{4.21}$$

for any $\mu \in P(S)$, *where* $I(f)$ *is given by* (4.20).

For example, we have the non-Gaussian limit

$$\lim_{T\to\infty} \frac{1}{T}\int_0^T X^*(t)dt = E\{X^*(0)\} = \gamma \in P(R^d) \text{ a.s.;} \tag{4.22}$$

i.e.,

$$\lim_{T\to\infty}\frac{1}{T}\int_0^T X^*(t, A)dt = \upsilon(A), \tag{4.23}$$

where $\upsilon(A)$ is the expected value of the equilibrium random measure on the set A. The collection $\{r_{k_1,\ldots,k_d}\}$ forms the joint moment system of the expected steady-state distribution of the random cluster centered at the empirical mean.

For the continuous limit $Y(t)$ of the multilevel branching process, the long-term extinction problem is solved in Dawson and Hochberg (1991) for various combinations of the values of c_i and γ_i (i = 1,2) using the martingale-problem formulation. In addition, the following generalizations of the growth rate and conditional limit law (conditioned on non-extinction of the process) for ordinary critical continuous-state branching are obtained for the continuous multilevel process $Y(t)$:

THEOREM 4.8. (a) *Let Y_t denote the multilevel branching continuous limit process with $\gamma_1 = \gamma_2 = 2$, $c_1 = c_2 = 0$. Then*

$$\lim_{t\to\infty} t^2 P\{\langle x, Y_t\rangle > 0\} = c, \text{ a constant.} \tag{4.24}$$

(b) *For $A \in B((0,\infty))$ and $t > 0$, let*

$$Y_K(t,A) := Y(Kt,\{x: x/K \in A\})/K.$$

Then conditioned on $\langle x,Y(t)\rangle > 0$ and $Y(0) = \delta_x$, $Y_K(t)$ converges weakly to a random measure with Laplace functional

$$L(\phi) = 1 - xV(t,\phi) \tag{4.25}$$

where

$$V(t,\phi) = (c_0 x)^{-1} \lim_{K\to\infty} Kv(t,x/K) \tag{4.26}$$

for a constant c_0 *and function $v(t,x)$ satisfying the partial differential equation*

$$\frac{\partial v}{\partial t} = x\frac{\partial^2 v}{\partial x^2} - v^2 \tag{4.27}$$

with initial value $v(0) = \phi$, $\phi \in C_K(0,\infty)$.

(c) *For each $k \in Z^+$ there exists $c_k \in R$, called the k-th* limiting cumulant *of the total mass process, such that (formally) as $t \to \infty$,*

$$\log E_{\delta_x}\left(exp\left[\frac{-\theta < x,Y(t) >}{t^2}\right] \middle| < x,Y(t) > \ > 0\right) \to \sum_{k=1}^{\infty} (-\theta^k)\, c_k/k!. \tag{4.28}$$

We note that a growth rate of t^2 in equations (4.24) and (4.28) for multilevel critical branching replaces the usual rate of t for ordinary critical branching.

In the two-level branching diffusion model, analysis of the dual process in [6] yields the following:

THEOREM 4.9. *Let $Y(t)$ denote the continuous limit of the two-level branching diffusion on R^d with spatial diffusion governed by a symmetric stable process of index α, and let the* total mass process $\{\bar{Y}(t) : t \geq 0\}$ *be defined by*

$$\bar{Y}(t) = \int \ \mu Y(t,d\mu). \tag{4.29}$$

Then, $E\{<\bar{Y}(t),\phi>\}$ remains bounded if and only if $d > 2\alpha$.

C. *Renormalization results.* The single-locus Fleming-Viot process $\{X(t): t \geq 0\}$ on R^d with $A = \Delta_d$ can be rescaled as follows. Let

$$X_\varepsilon(t,A) = X(t/\varepsilon^2, \{x{:}\varepsilon x \in A\}). \tag{4.30}$$

Then, $\{X_\varepsilon(t)\}$ has generator L_ε which is related via

$$L_\varepsilon F_f(\mu) = L_\varepsilon^d F_\mu(f) \tag{4.31}$$

to a function-valued dual process $\{Y_\varepsilon(t)\}$ with generator $L_\varepsilon{}^d$ given by

$$L_\varepsilon^d F_\mu(f) = F_\mu(\Delta_{N(f)} f) + \frac{c}{\varepsilon^2} \sum_{j=1}^{N(f)} \sum_{\substack{k=1 \\ j \neq k}}^{N(f)} [F_\mu(\phi_{jk} f) - F_\mu(f)]\,, \tag{4.32}$$

where Φ_{jk} is the operator defined in section B. For $f \in C(S^N)$,

$$\lim_{\varepsilon \to 0} E_{\delta_x}\{F_f(X_\varepsilon(t))\} = \lim_{\varepsilon \to 0} E_f\{F_{\delta_x}(Y_\varepsilon(t))\} \tag{4.33}$$

$$= H_t^1(\Psi^1 f)(x) = E\{F_f(\delta_{W(t)})\}\,,$$

where $H_t^N f(x)$ is defined in (4.15), $\{W(t): t \geq 0\}$ denotes the standard Wiener process in R^d, and $\psi^1 f(x) \equiv f(x,x,...,x)$. We have thus proven the following:

THEOREM 4.10. *Let $\{X(t): t \geq 0\}$ be the Fleming-Viot process in R^d with diffusion generator $A = \Delta_d$. Assume that $X(0)$ has compact support. Then, the fi-*

nite-dimensional distributions of the rescaled process $\{X_\varepsilon(t): t \geq 0\}$ *defined by* (4.30) *converge to those of the probability-measure-valued process which consists of a single unit atom undergoing Brownian motion in* R^d.

In [2], this scaling limit is obtained, with convergence in the sense of weak convergence of probability measures on $\Omega_c^{P(S)}$, using martingale techniques. Vaillancourt (1990a, b) has similarly derived scaling limit results for a class of weakly-interacting Fleming-Viot processes.

References

[1] Dawson, D.A. (1975). Stochastic evolution equations and related measure processes. *J. Multivariate Analysis* **5** 1-52.

[2] Dawson, D.A. (1977). The critical measure diffusion process. *Z. Wahrsch. Verw .Geb.* **40** 125-145.

[3] Dawson, D.A. and Hochberg, K.J. (1979). The carrying dimension of a stochastic measure diffusion. *Ann. Probability* **7** 693-703.

[4] Dawson, D.A. and Hochberg, K.J. (1982). Wandering random measures in the Fleming-Viot model. *Ann. Probability* **10** 554-580.

[5] Dawson, D.A. and Hochberg, K.J. (1983). Qualitative behavior of a selectively neutral allelic model. *Theoretical Population Biology* **23** 1-18.

[6] Dawson, D.A. and Hochberg, K.J. (1985). Function-valued duals for measure-valued processes and applications. *Particle Systems, Random Media, and Large Deviations*, R. Durrett, ed., *Contemporary Mathematics* **41** 55-69.

[7] Dawson, D.A. and Hochberg, K.J. (1991). A multilevel branching model. *Adv. Appl. Prob.* In press.

[8] Dawson, D.A., Hochberg, K.J. and Wu, Y. (1990). Multilevel branching systems. *White Noise Analysis -- Mathematics and Applications*, T. Hida, H.H. Kuo, J. Potthoff, and L. Streit, eds., World Scientific Pub; Singapore - New Jersey - London, 93-107.

[9] Dawson, D.A., Iscoe, I. and Perkins, E.A. (1989). Super-Brownian motion: path properties and hitting probabilities. *Probab. Th. Rel. Fields* **83** 135-205.

[10] Dawson, D.A. and Kurtz, T.G. (1982). Application of duality to measure-valued diffusion processes. *Advances in Filtering and Optimal Stochastic Control. Lecture Notes in Control and Information Sciences*, Springer-Verlag, Berlin-Heidelberg-New York, **42** 91-105.

[11] Dynkin, E.B. (1988). Representation for functionals of superprocesses by multiple stochastic integrals, with applications to self-intersection local times. *Asterisque* **157-158** 147-171.

[12] Dynkin, E.B. (1989). Superprocesses and their linear functionals. *Trans. Am. Math. Soc.* **314** 255-282.

[13] Dynkin, E.B. (1989). Regular transition functions and regular superprocesses. *Trans. Am. Math. Soc.* **316** 623-634.

[14] Dynkin, E.B. (1989). Branching particle systems and superprocesses. Preprint.

[15] Ethier, S.N. and Griffiths, R.C. (1987). The infinitely-many-sites model as a measure-valued difffusion. *Ann. Probability* **15** 515-545.

[16] Ethier, S.N. and Griffiths, R.C. (1990). The neutral two-locus model as a measure-valued difffusion. *Adv. Appl. Prob.* **22** 773-786.

[17] Ethier, S.N. and Kurtz, T.G. (1987). The infinitely-many-alleles model with selection as a measure-valued diffusion. *Stochastic Methods in Biology. Lecture Notes in Biomathematics*, Springer-Verlag, Berlin-Heidelberg-N.Y., **70** 72-86.

[18] Fleming, W.H. and Viot , M. (1979). Some measure-valued Markov processes in population genetics theory. *Indiana Univ. Math. J.* **28** 817-843.

[19] Hochberg, K.J. (1980). Medidas de difusion estocasticas. *Conferencias Sobre Sistemas Estocasticos*, Centro de Investigacion del Inst. Politecnico Nac., Mexico, 97-107.

[20] Hochberg, K.J. (1983). Microscopic clustering of population processes. Population Biology. *Lecture Notes in Biomathematics*, Springer-Verlag, Berlin-Heidelberg- N.Y., **52** 18-24.

[21] Hochberg, K.J. (1986). Stochastic population theory: mathematical evolution of a genetical model. *The Merging of Disciplines: New Directions in Pure, Applied, and Computational Mathematics*, R.E. Ewing, K.I. Gross and C.F. Martin, eds., Springer-Verlag, N.Y., 101-115.

[22] Iscoe, I. (1986). A weighted occupation time for a class of measure-valued critical branching Brownian motion. *Probab. Th. Rel. Fields* **71** 85-116.

[23] Iscoe, I. (1986). Ergodic theory and a local occupation time for measure-valued branching processes. *Stochastics* **18** 197-243.

[24] Iscoe, I. (1988). On the supports of measure-valued critical branching Brownian motion. *Ann. Probab.* **16** 200-221.

[25] Perkins, E.A. (1988). A space-time property of a class of measure-valued branching diffusions. *Trans. Am. Math. Soc.* **305** 743-795.

[26] Perkins, E.A. (1989). The Hausdorff measure of the closed support of super Brownian motion. *Ann. Inst. H. Poincare* **25** 205-224.

[27] Perkins, E.A. (1990). Polar sets and multiple points for super Brownian motion. *Ann. Probab.* **18** 453 - 491.

[28] Stroock, D.W. and Varadhan, S.R.S. (1979). *Multidimensional Diffusion Processes*. Springer-Verlag, Berlin-Heidelberg-New York.

[29] Vaillancourt, J. (1990a). Interacting Fleming-Viot processes. *Stoch. Proc. Appl.* **36** 45-57.

[30] Vaillancourt, J. (1990b). On the scaling theorem for interacting Fleming-Viot processes. *Stoch. Proc. Appl.* **36** 263-267.

[31] Zähle, U. (1988). Self-similar random measures, I. Notion, carrying Hausdorff dimension, and hyperbolic distribution. *Probab. Th. Rel. Fields* **80** 79-100.

SADDLEPOINT APPROXIMATIONS IN THE CASE OF INTRACTABLE CUMULANT GENERATING FUNCTIONS

John E. Kolassa
IBM T. J. Watson Research Center

Abstract

Saddlepoint Approximations have long been used to approximate densities and distribution functions of random variables with known cumulant generating function defined on an open interval about the origin. This approximation has very desirable asymptotic properties when approximating densities and tail probabilities for sums of random variables, and also often performs remarkably well for small sample sizes, including samples of one.

Calculating the saddlepoint approximation requires calculating the Legendre transform of the log of the cumulant generating function. In some cases this cumulant generating function may be unavailable; in other cases the Legendre transform is difficult to calculate analytically. This paper discusses modifications to the saddlepoint approximation necessary when the cumulant generating function is replaced by a similar but more tractable function whose Legendre transform can be given explicitly. Calculations for the logistic distribution are presented to illustrate the case of a known but intractable cumulant generating function, and an example involving an overdispersed binomial model is presented to illustrate the case of an unavailable cumulant generating function. An application to a random effects logistic linear model is discussed.

1. The Problem. Consider the following problem: $X_1, X_2, X_3, ..., X_n, ...$ are independent and identically distributed random variables. Assume that their common distribution is continuous, with probability density function $f_1(x)$ and cumulant generating function $K(t) = \log E[e^{tX}]$, defined on some convex set $I \subset \boldsymbol{R}$. Without loss of generality assume further that $E[X_i] = 0$. I wish to approximate the density function $f_n(s)$ and distribution function $F_n(s)$ of

$$S = \frac{X_1 + ... + X_n}{\sqrt{n}}$$

analytically, and am faced with three options:

a. Convolute the density n times.

b. Use the classical Edgeworth series approximation to $f_n(x)$.

c. Use a saddle point approximation to $f_n(x)$.

Option (a) is often too difficult. The Edgeworth series of option (b) often behaves poorly in the tails. In order to exercise option (c), one must solve the saddlepoint equation

$$\sqrt{n}K'(t_e/\sqrt{n}) = s \tag{1}$$

The solution t_e is known as the saddlepoint. The value t_e is that element of the domain of K for which the tilted density, described in section 3, has its mean exactly at s. Often this is easy to solve analytically, but in some statistical calculations this is very difficult (Table 1). In these situations I propose finding a similar but computationally simpler approximate cumulant generating function L, performing the computations with the saddlepoint of L rather than that of K, and substituting it into the derivatives of K as specified in the standard saddlepoint approximation. Below I discuss the modifications to this process needed to retain good asymptotic properties. Denote the difference between these cumulant generating functions $K(t) - L(t)$ by $a(t)$. Suppose furthermore that

i. 0 is in the interior of I, the domain of K.
ii. L is analytic on $I \times i\mathbf{R}$
iii. $L(0) = 0$ and $L'(0) = 0$.
iv. a' is bounded on I.
v. $K(t) - tL'(t) \leq 0$ for all $t \geq 0$.

Table 1: *Examples of Legendre-Fenchel Calculations*

Distribution	Density	Domain	Cumulant Generating Function	Domain	Saddlepoint
Normal	$\frac{\exp\left[-\frac{1}{2}x^2\right]}{\sqrt{2\pi}}$	R	$\frac{1}{2}t^2$	R	x
Uniform	1	$\left[-\frac{1}{2}, \frac{1}{2}\right]$	$\log\left[\frac{\sinh\left(\frac{1}{2}t\right)}{t/2}\right]$	R	intractable
Exponential	$\exp[-x]$	$[0, \infty)$	$-\log[1 - t]$	$(-\infty, 1)$	$(x - 1)/x$
Double Exponential	$\frac{1}{2}\exp[\lvert x\rvert]$	R	$-\log[1 - t^2]$	$(-1, 1)$	$(\sqrt{1+x^2} - 1)/x$
Logistic	$\frac{\exp[-x]}{(1+\exp[-x])^2}$	R	$\log\left[\frac{\pi t}{\sin(\pi t)}\right]$	$(-1, 1)$	intractable

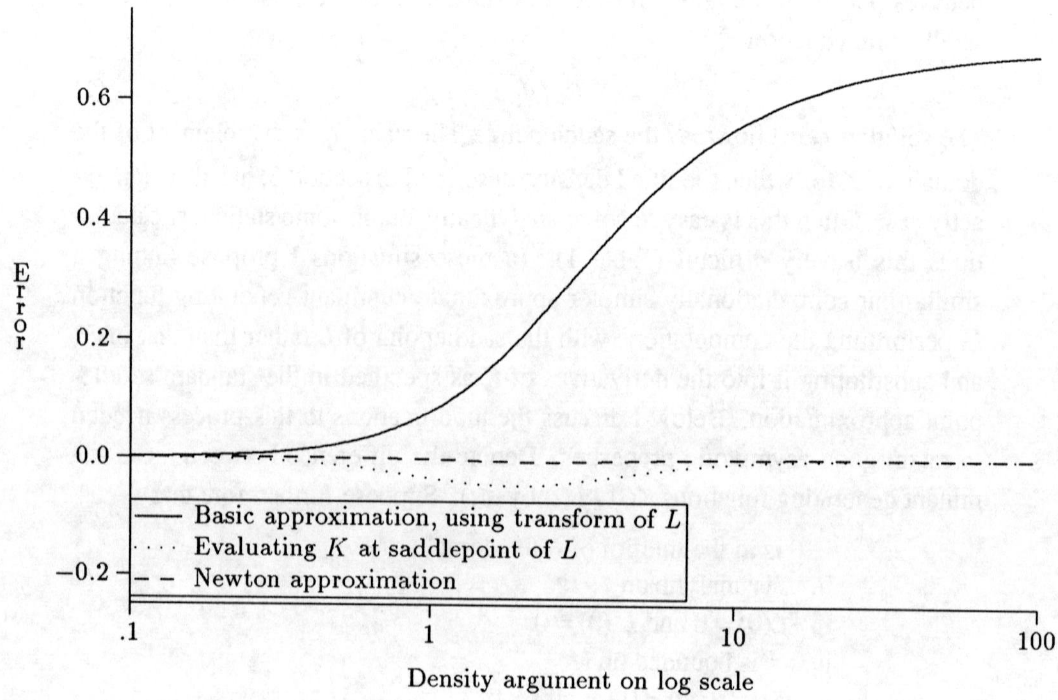

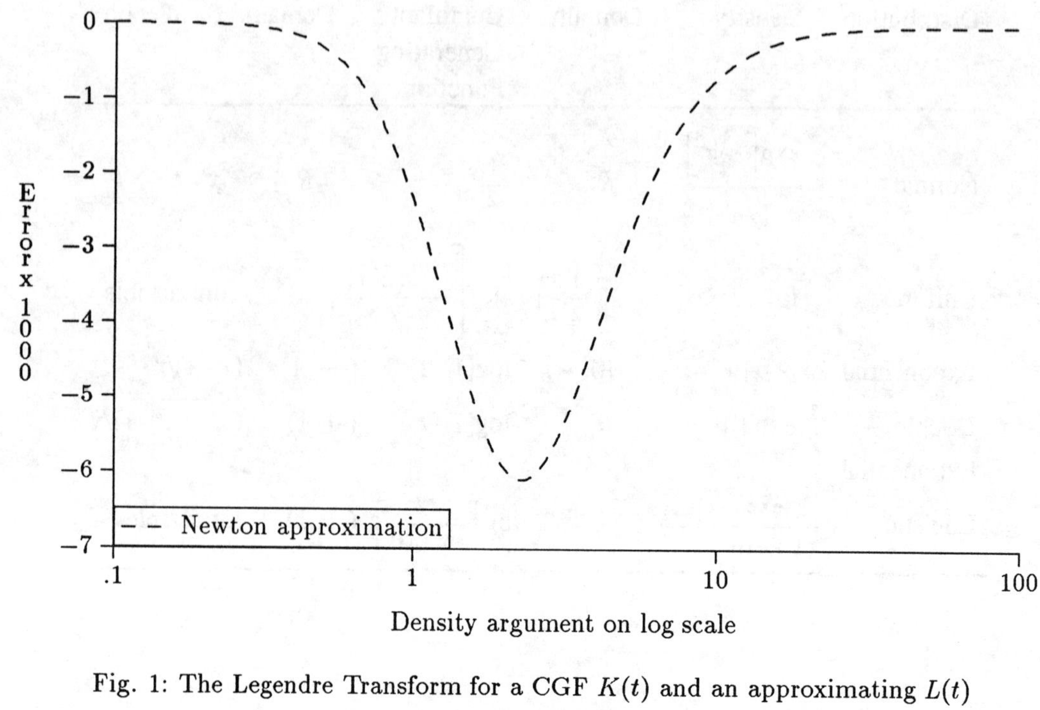

Fig. 1: The Legendre Transform for a CGF $K(t)$ and an approximating $L(t)$

In particular, the cumulant generating functions for the logistic and double exponential distributions satisfy these requirements. Note that although the saddlepoint equation $\sqrt{n}K'(t_e/\sqrt{n}) = s$ need not have a solution for all s [Daniels (1954)], condition (i) implies that a solution always exists for large enough n.

One might try to substitute the approximate saddlepoint t_a for the exact saddlepoint t_e in the expression for the saddlepoint approximation to f_n. (Figure 1). In fact, using knowledge about a, one can improve on this naive approximation.

2. The Edgeworth Series. Let the random variables X_i and S_n be defined as in Section 1. The cumulant generating function of S_n is $K_n(t) = \sqrt{n}K(t/\sqrt{n})$. Note that these functions are defined for at least t pure imaginary. In the present case, however, we assume that K exists on $I + i\mathbf{R}$ for some open interval I containing 0. One may easily show that K is infinitely differentiable at every point in this domain. Define the cumulant of order r of X_i to be $\kappa_r = K^{(r)}(0)$. Then the cumulant of order r of S_n is $\kappa_r^n = \kappa_r n^{1-r/2}$.

The density $f_n(s)$ and the cumulative distribution function $F_n(s)$ can, for large n, be well-approximated for moderate values of s, by an Edgeworth series. This series has the form

$$e_m(s;\kappa^n) = \phi(s)\left\{\sum_{k=2}^{m}\sum_{j=0}^{3(k-2)}\gamma_{j,k}h_j(s)\right\},$$

where the standard Hermite polynomials $h_r(s)$ are given by $(-1)^r\phi^{(r)}(s)/\phi(s)$. When $m = 2$, this reduces to the usual normal approximation, and $\gamma_{0,2} = 1$ and $h_0 \equiv 1$. The function ϕ is the standard normal density, and differentiation is with respect to s. The coefficients $\gamma_{j,k}$ are sums of products of the first k cumulants. McCullagh (1987, Chapter 5) gives a thorough description of this series. Feller (1971, page 535) shows that under Cramer's condition,

$$\lim_{|t|\to\infty}\sup|\exp[K(it)]| < 1,$$

that

$$f_n(t) = e_r(t;\kappa^n) + o\left(n^{-\frac{r-2}{2}}\right),$$

holds uniformly in t. Cramer's condition is always satisfied in cases where the random variables X_i have a density.

The Edgeworth series for the density can easily be integrated analytically to give an approximation $E_r(t;\kappa^n)$ to the cumulative distribution function, valid to the same order.

3. The Expansion for the Density. Denote the density of S by $f_n(s)$. Embed the density $f_n(s)$ in the exponential family $\{f_n(s;t) = \exp[ts - nK(t/\sqrt{n})] f_n(s)\colon t \in I_n\}$, where I is the domain of the cumulant generating function $K(t)$ of X_1, and $I_n = \sqrt{n}I$ is the domain of the cumulant generating function $K_n(t) = nK(t/\sqrt{n})$ of S_n. Note that $f_n(s;t)$ has the cumulant generating function

$$\hat{t} \rightarrow n[K((\hat{t}+t)/\sqrt{n}) - K(t/\sqrt{n})].$$

Standard saddlepoint techniques proceed by choosing t_e to satisfy (1); here assume that a solution to this equation is not available, but that a suitable approximation L to K exists, with properties as described in section 1, and such that one can solve

$L'(t_a/\sqrt{n}) = s/\sqrt{n}$. In what follows, t_a is treated as an approximation to t_e, and is substituted into K in place of t_e. Corrections for the difference between t_a and t_e then enter the approximation through the function N of t_a described below.

By construction,

$$\begin{aligned} f_n(s) &= \exp[nK(t_a/\sqrt{n}) - t_a s] f_n(s;t_a) \\ &= \exp[nK(t_a/\sqrt{n}) - t_a s] f_n(\mu_a + \sigma_a N(t_a);t_a) \end{aligned}$$

where

$$\mu_a = \sqrt{n}K'(t_a/\sqrt{n}),$$

$$\sigma_a = \sqrt{K''(t_a\sqrt{n})}, \tag{2}$$

and

$$N(t_a) = (s - \sqrt{n}K'(t^a/\sqrt{n}))/\sigma_a = \sqrt{n}(L'(t_a/\sqrt{n}) - K'(t_a/\sqrt{n}))/\sigma_a$$

Approximate the density $\sigma_a f_n(\mu_a + \sigma_a N;t_a)$ as an Edgeworth series

$$e_m(N;\beta^n) = \phi(N)\left\{\sum_{k=2}^{m}\sum_{j=0}^{3(k-2)}\gamma_{j,k}h_j(N)\right\},$$

where

$$\beta_r^{n,t_a} = n^{1-r/2}K^{(r)}(t_a/\sqrt{n})/[K^{(2)}(t_a/\sqrt{n})]^{n/2}$$

for $2 \leq r \leq m$, and $\beta_n^{n,t_a} = 0$. Let β^{n,t_a} denote the collection of these. Denote the error in this approximation by $R_n^m(s)$. The β^{n,t_a} are the cumulants of a variable with the density $f_n(s;t_a)$, after normalizing to unit variance. The coefficients γ are the appropriate coefficients for an Edgeworth series using cumulants β^{n,t_a}. Then

$$f_n(s) = \exp[nK(t_a/\sqrt{n}) - t_a s]\frac{1}{\sigma_a}[e_m(N;\beta^n) + R_n^m(s)] \tag{3}$$

The error term is of order $O(n^{(1-m)/2})$. The size of this term depends on the cumulants β^{n,t_a}, which depend in turn on s. Daniels (1954) describes conditions on the density of the X_i which in turn imply that β_r^{n,t_a} are of order $O(n^{1-m/2})$ uniformly in t_a. The appendix of this paper provides a proof that in these cases, the above bound on the error term is uniform.

In the case where the saddlepoint equation (1) is solvable, then $K = L$, $t_a = t_e$, and N is identically 0. Then the expression (3) simplifies to the usual saddlepoint density, involving the evaluation of an Edgeworth series at 0. Note that in an Edgeworth series for a density, all odd powers of $1/\sqrt{n}$ are multiplied by odd-ordered Hermite polynomials, which are 0 when evaluated at 0. Hence, an Edgeworth series evaluated at 0 has only terms in integer powers of $1/n$. The leading term of this approximation,

$$[2\pi K^{(2)}(t_e/\sqrt{n})]^{-1/2}\exp[nK(t_e/\sqrt{n}) - t_e s], \tag{4}$$

is accurate to order $O(1/n)$ rather than $O(1/\sqrt{n})$. Note also that this first-order approximation is always positive. Daniels (1954) provides an early account of these developments.

4. The Expansion for the Tail Probabilities. I now desire an approximation to tail probabilities $Q_n(s) = P[S_n \geq s]$ for the random variable S_n described earlier,and proceed as follows:

Denote the density of S by $f_n(s)$. Again embed the density $f_n(s)$ in the exponential family $\{f_n(s;t) = \exp[ts - nK(t/\sqrt{n})]f_n(s) : t \in I_n\}$, and chose t_a such that $\sqrt{n}L'(t_a/\sqrt{n}) = s$. Let $z_a = t_a\sqrt{K''(t_a\ \sqrt{n})}$. Then,

$$Q_n(s) = \int_s^\infty f_n(y)\,dy$$

$$= \exp[nK(t_a/\sqrt{n}) - t_a\sqrt{n}K'(t_a/\sqrt{n})]\int_s^\infty \exp[-t_a(y - \sqrt{n}K'(t_a/\sqrt{n}))]$$

$f_n(y;t_a)dy$

$$= \exp[nK(t_a/\sqrt{n}) - t_a\sqrt{n}K'(t_a/\sqrt{n})]\int_{-N(t_a)}^\infty \exp[-z_a\xi]\,\sigma_a f_n(\mu_a + \sigma_a\xi;t_a)\,d\xi$$

where

$$\xi = (y - \sqrt{n}K'(t_a/\sqrt{n}))/\sigma_a,$$

σ_a, μ_a, and N are defined as in (2). Approximate the density $\sigma_a f_n(\mu_a + \sigma_a\xi;t_a)$ as an Edgeworth series

$$e_m(\xi;\beta^n) = \phi(\xi)\left\{\sum_{k=2}^{m}\sum_{j=0}^{3(k-2)} \gamma_{j,k}h_j(\xi)\right\},$$

where the β^{n,t_a} are as before. Then

$$Q(s) = \exp[nK(t_a/\sqrt{n}) - t_a\sqrt{n}K'(t_a/\sqrt{n})]\int_{-N(t_a)}^\infty \exp[-z_a\xi]e_m(\xi;\beta^n)d\xi + R_n^m(s)$$

where the remainder term $R_n^m(s)$ will be discussed later,

$$= \exp[nK(t_a/\sqrt{n}) - t_a\sqrt{n}K'(t_a/\sqrt{n})]\sum_{k=2}^{m}\sum_{j=0}^{3(k-2)} \gamma_{j,k}I_j(z_a, -N(t_a)) + R_n^m(s), \quad (5)$$

where $I_j(a, b) = \int_b^\infty \exp[-a\xi]\,\phi(\xi)\,h_j(\xi)\,d\xi$. Integration by parts shows that

$$I_0(a,b) = \exp\left[\frac{1}{2}a^2\right][1 - \Phi(a+b)]$$

$$I_j(a,b) = aI_{j-1}(a,b) - \exp[-ab]\,\phi(b)\,h_{j-1}(\xi),$$

whence by induction we find that

$$I_j(a,b) = \exp\left[\frac{1}{2}a^2\right]\{(-a)^j[1 - \Phi(a+b)] + \phi(a+b)\sum_{l=0}^{j-1}(-a)^l h_{j-1-l}(b)\}.$$

Hence we can in principle evaluate this series to any order desired; however, calculations beyond the first order approximation are very messy. The second appendix contains a proof that the error $R_n^m(s)$ is of order $O\left(n^{\frac{1-m}{2}}\right)$ uniformly in s. The approximation in the case when $m = 2$, accurate to order $O(1/\sqrt{n})$, is

$$\exp[nK(t_a/\sqrt{n}) - t_a\sqrt{n}K'(t_a/\sqrt{n}) + \frac{1}{2}(t_a)^2K''(t_a/\sqrt{n})\}][1 - \Phi(z_a - N(t_a))]. \quad (6)$$

Here $z_a = t_a\sqrt{K''(t_a\ \sqrt{n})}$. This is the main result of this paper.

Note that if $m = 2$, the Berry-Esseen theorem [Feller (1971), p. 542] implies that the result (5) does not depend on X_i having a density, but only on the existence of a finite third moment, which in turn is guaranteed by condition i of section 1. Hence the result (6) holds whether or not the X_i have a density.

Note also that the difference between $nK(t_a/\sqrt{n}) - t_a\sqrt{n}K'(t_a/\sqrt{n}) + \frac{1}{2}(t_a)^2K''({}^tt_a/\sqrt{n})$ and this quantity with K replaced by L is $na(t_a/\sqrt{n}) - t_a\sqrt{n}a'(t_a) + \frac{1}{2}(t_a)^2a''(t_a) = (t_a/\sqrt{n})^3a^{(3)}(\theta)/6$, for some $\theta \in [0, t_a)$, through a judicious use of Taylor's theorem. If the third derivatives of a is uniformly bounded, then, this difference is of order $O(1/\sqrt{n})$. The same argument shows that substituting L'' for K'' in the definitions of N and z_a introduces an error of order $O(1/\sqrt{n})$. Hence (6) can be changed to

$$\exp[nL(t_a/\sqrt{n}) - t_a s + \frac{1}{2}(t_a)^2L''(t_a/\sqrt{n})\,[1 - \Phi(z_a - N(t_a))] \quad (7)$$

while remaining accurate to $O(1/\sqrt{n})$ uniformly in s, and once again the requirement that the X_i have a density can be dropped.

Note that in the case where the saddlepoint equation (1) is solvable, and hence $K = L$, $t_a = t_e$, and N is identically 0, expression (6) agrees with the usual saddlepoint approximation given by Robinson (1982),

$$\exp[nK(t_e/\sqrt{n}) - t_e s + \frac{1}{2}z_m^2]\left\{[1 - \Phi(z_m)]\left(1 - \frac{\beta_3^{n,t_e}z_m^3}{6}\right) + \phi(z_m)\frac{\beta_3^{n,t_e}}{6}(z_m^2 - 1)\right\}, \quad (8)$$

to order $O(1/\sqrt{n})$. Note that the coefficient β_3^{n,t_e} has a factor of $1/\sqrt{n}$ included. Daniels (1987) discusses alternate derivations for this expansion, as well as other expansions for tail probabilities. These in turn might inspire similar variations on the expansion presented here.

5. The Choice of L. For the logistic examples considered in this paper, the cumulant generating functions are defined on an open interval (τ_1, τ_2), with $\tau_1 < 0 < \tau_2$, and there exist r_1 and r_2 such that $(\tau_i - t) K'(t) \to r_i$ as $t \to \tau_i$, for $i = 1$ and 2. Then the approximate cumulant generating function $L(t) = -\Sigma_{i=1}^{2} [r_i \log(1 - t/\tau_i) + (r_i t)/\tau_i]$ will satisfy the requirements i-v of section 1, and the resulting approximate saddlepoint t_a is the solution of the quadratic equation in t:

$$\frac{r_1}{\tau_1 - t} + \frac{r_2}{\tau_2 - t} - \frac{r_1}{\tau_1} - \frac{r_2}{\tau_2} = s.$$

Finding approximate cumulant generating functions more closely fitting the actual functions appears to be very difficult.

6. An Example: The Logistic Distribution. Consider for example the logistic distribution, whose density is $\dfrac{\exp[-x]}{(1+\exp[-x])^2}$ and whose cumulant generating function is

$$K(t) = \log\left(\frac{\pi t}{\sin(\pi t)}\right) \text{ for } t \in (-1,1),$$

and set $n = 1$, so that $S = X_1$. The approximating cumulant generating function defined by the algorithm described in section 5 is

$$L(t) = -\log(1 - t^2) \text{ for } t \in (-1,1),$$

the cumulant generating function of the double exponential, with density $\frac{1}{2}\exp[-|x|]$, defined over the same interval. The saddlepoint of L is easily found to be $(\sqrt{1+x^2} - 1/x)$. Straight-forward calculations verify the conditions (2) are fulfilled; the fifth condition is checked by examining coefficients of the series expansion for $K(t)$ - $tL'(t)$ and noting that they are all negative. The function L was chosen so that the difference $a(t)$ between $K(t)$ and $L(t)$ is bounded, and has a bounded first derivative.

I plotted the differences between the logarithms of various approximating densities and the logarithm of the true density (Figure 2). The approximations considered are the Edgeworth series, the Saddle Point series in which the approximating saddlepoint t_a satisfying $L'(t_a) = x$ is used naively instead of t_e satisfying $K'(t_e) = x$, the quasi-saddlepoint approximation (6), and the true saddle point series. The Edgeworth series performs best for small values of the ordinate, but for large values this approximation fails miserably. The other three approximations all perform similarly. Maximal fluctuation in the differences from the logarithm of the true density are about the same, implying that under an optimal standardization, these approximations would have similar maximal errors.

Part of the discrepancy between the approximation and the exact saddlepoint approximation can be explained by the difference in variance standardizations used. The true saddle point approximation uses the second derivative of the cumulant generating function evaluated at the true saddle point, while the approximation uses the this derivative evaluated at the saddle point of the simpler cumulant generating function. The ratio of these standardizations is far from unity when the quasi-saddlepoint approximation and the exact saddle point approximation are far apart.

The quasi-saddlepoint approximation may be improved by using a one-term Taylor series approximation in $t_e - t_a$ to $K''(t_e)$. Note that by Taylor's theorem, $K'_n(t_e) \approx L'_n(t_a) + a'_n(t_a) + (t_e - t_a)K''_n(t_a)$, and hence $t_e - t_a \approx -a'_n(t_a)/K''_n(t_a)$. Hence $K''_n(t_a) \approx K''_n(t_a) + (t_e - t_a)K'''_n(t_a) \approx K''_n(t_a) - a'_n(t_a)K'''_n(t_a)/K''_n(t_a)$.

Replacing $\sqrt{K''_n(t_a)}$ in the definition of σ_a by this expression results in a variance standardization much closer to that of the exact saddlepoint, and a resulting approximation that works better than the exact saddlepoint approximation (Figure 3).

Note that the choice of approximating cumulant generating function was crucial here. For example, if L is modified by multiplying by the ratio of the variances corresponding to K and L, in order to match variances, the resulting difference is not well-behaved, and indeed the quasi-saddlepoint approximation works poorly (Figure 4).

7. An Application. Convolution of the normal and logistic distributions has important applications in random effects logistic models where the random effect is normally distributed. Consider the following logistic model for random variables X_i taking on the values 0 or 1. The probability of realizing a 1 depends on the random effect ε_i:

$$\text{logit } P(X_i = 1 | \varepsilon_i) = \mu + \varepsilon_i$$

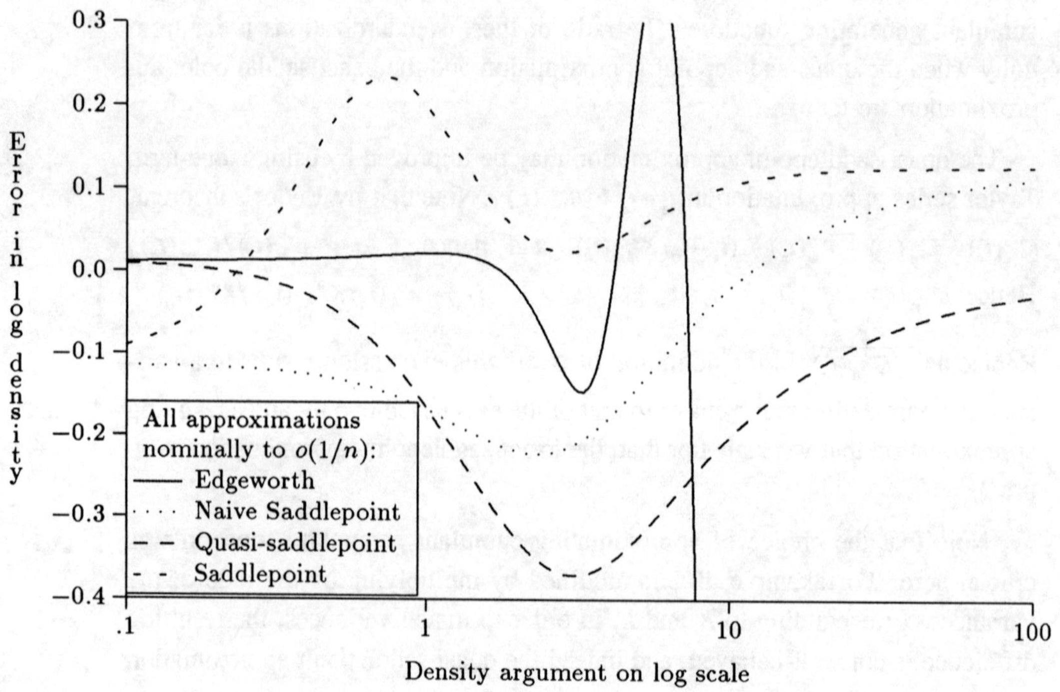

Fig. 2: Error in Unstandardized Approximations to log of Logistic Density

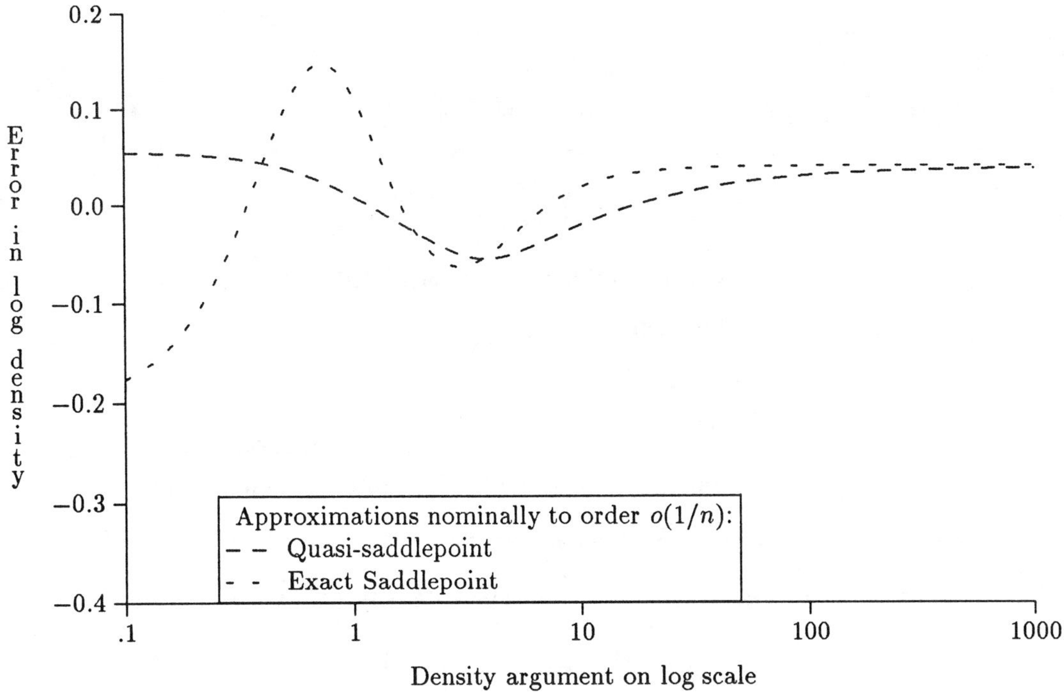

Fig. 3: Error in Approximation with Improved Variance Standardization

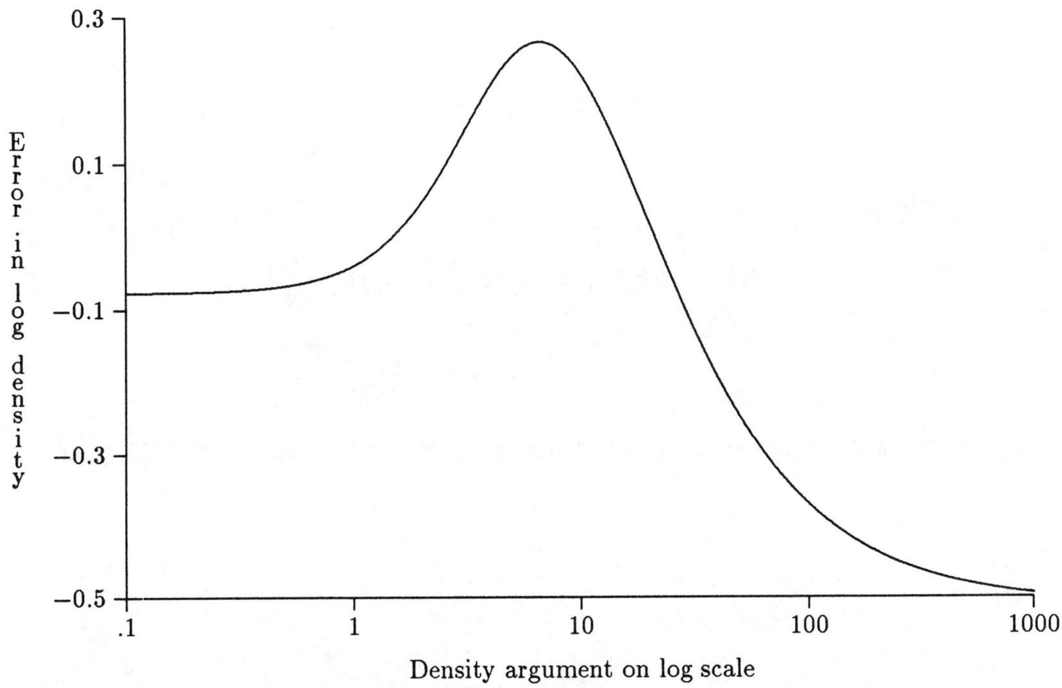

Fig. 4: Performance of Quasi-saddlepoint Approximation for Poorly Chosen L

where ε_i are normals with zero mean and variance σ^2. I wish to derive approximations to the unconditional means and variances of the $P(X_i = 1|\varepsilon_i)$'s.

Note that

$$E(P(X_i = 1|\varepsilon_i)) = \int_{-\infty}^{\infty} P(X_i = 1|\varepsilon_i)\frac{1}{\sigma}\phi\left(\frac{\varepsilon_i}{\sigma}\right)d\varepsilon_i$$

$$= \int_{-\infty}^{\infty} G_1(\mu - \varepsilon)\frac{1}{\sigma}\phi\left(\frac{\varepsilon}{\sigma}\right)d\varepsilon$$

where $G_1(x)$ is the cumulative distribution function $e^x/(1+e^x)$.

$$= G_2(\mu)$$

where $G_2(x)$ is the cumulative distribution function of the sum of a logistic random variable and an independent $N(0, \sigma^2)$ random variable.

Hence we see that $E(P(X_i = 1|\varepsilon_i))$ can be approximated by (6), where

$$K(t) = \log(\pi t \csc(\pi t)) + \frac{1}{2}t^2\sigma^2$$

$$L(t) = -\log(1 - t^2)$$

$$a(t) = \log(\pi t(1 - t^2)\csc(\pi t)) + \frac{1}{2}t^2\sigma^2$$

$$t_a = \frac{\sqrt{1 + x^2} - 1}{x}$$

Similarly,

$$E(P(X_i = 1|\varepsilon_i)^2) = \int_{-\infty}^{\infty} P(X_i = 1|\varepsilon_i)^2\frac{1}{\sigma}\phi\left(\frac{\varepsilon_i}{\sigma}\right)d\varepsilon_i$$

$$= \int_{-\infty}^{\infty} G_1(\mu - \varepsilon)^2\frac{1}{\sigma}\varphi\left(\frac{\varepsilon}{\sigma}\right)d\varepsilon.$$

Note that G_1^2 is an increasing differentiable function taking values in [0, 1], such that

$$\lim_{z \to -\infty} G_1^2(z) = 0$$

and

$$\lim_{z \to \infty} G_1^2(z) = 1$$

and hence is a distribution function. Differentiation shows that its density is

$$(2e^{2x}) / (1+e^x).$$

Integration shows that its cumulant generating function is $K(t) = \log(t(1+t)\pi\csc(\pi t))$ defined on (-2, 1). It also has first moment 1. Hence

$$E(P_i^2) = G_3(\mu - 1).$$

where G_3 is the cumulative distribution function corresponding to the cumulant generating function $K(t) = \log(t(1+t)\pi\csc(\pi t)) - t + \sigma^2 t^2/2$ defined on (-2, 1). The methods of section 5 lead to the approximation $L(t) = -\log((1-t)(2+t)) - t/2 + \log(2)$, also defined on the same interval, and satisfying the conditions (2). Furthermore, the saddlepoint equation arising from L is easy to solve. Hence we see that $E(P_i^2)$ can be approximated by (6), where K and L are as just described, and

$$a(t) = \log(\pi t(1-t^2)(2+t)\csc(\pi t)) + \frac{1}{2}t^2\sigma^2 + \frac{1}{2}t - \log(2)$$

$$t_a = \frac{\sqrt{1+\frac{9}{4}\left(x+\frac{1}{2}\right)^2} - 1}{x+\frac{1}{2}} - \frac{1}{2}.$$

These approximations appear at first glance to work well (Figures 5 and 6) but note approximation (6) performs poorly when evaluated at values of the ordinate less than the mean of the distribution. These approximations can easily be modified to calculate the lower tail when the ordinate is less than the mean. Taking care to calculate the correct tail, and subtracting from 1 if necessary, these approximations appear to work well over all, but for moderate σ, do not perform well enough to approximate the variance (Figure 7). Furthermore, the resulting discrepancy is more a result of the inability of the quasi-saddlepoint expression to approximate the true saddlepoint than the error inherent in the saddlepoint approximation.

8. A Second Example: An Overdispersed Binomial Distribution. Consider random variables X_i distributed as Bernoulli trials each with success probability p_i. I aim to calculate tail probabilities for the standardized sum S_n of

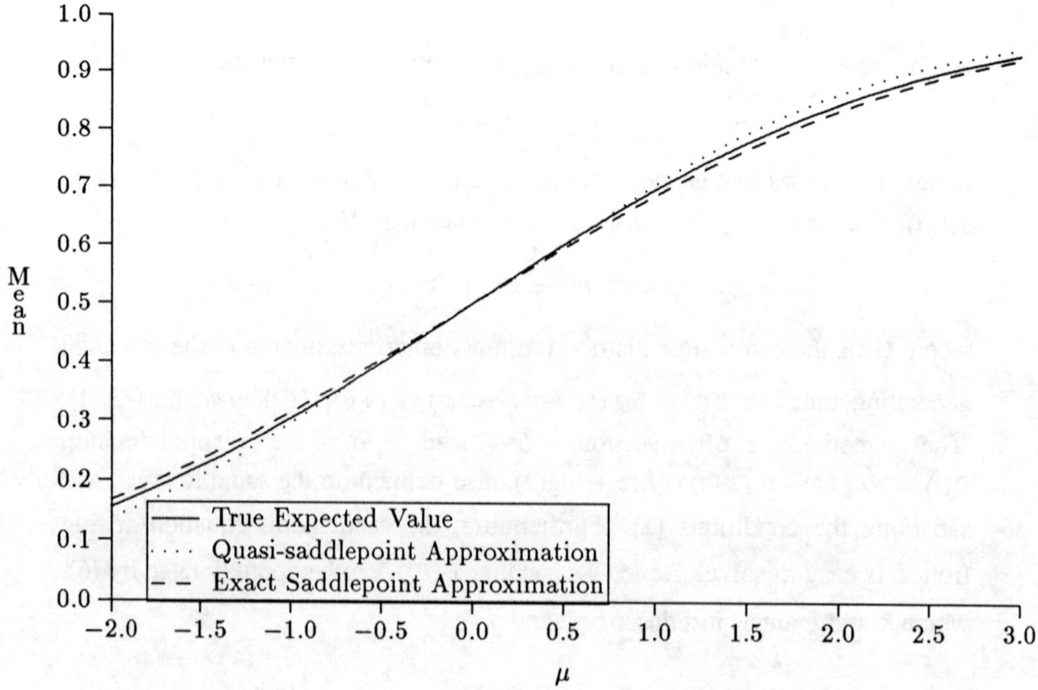

Fig. 5: Approximation to the First Moment of the Resulting Parameter

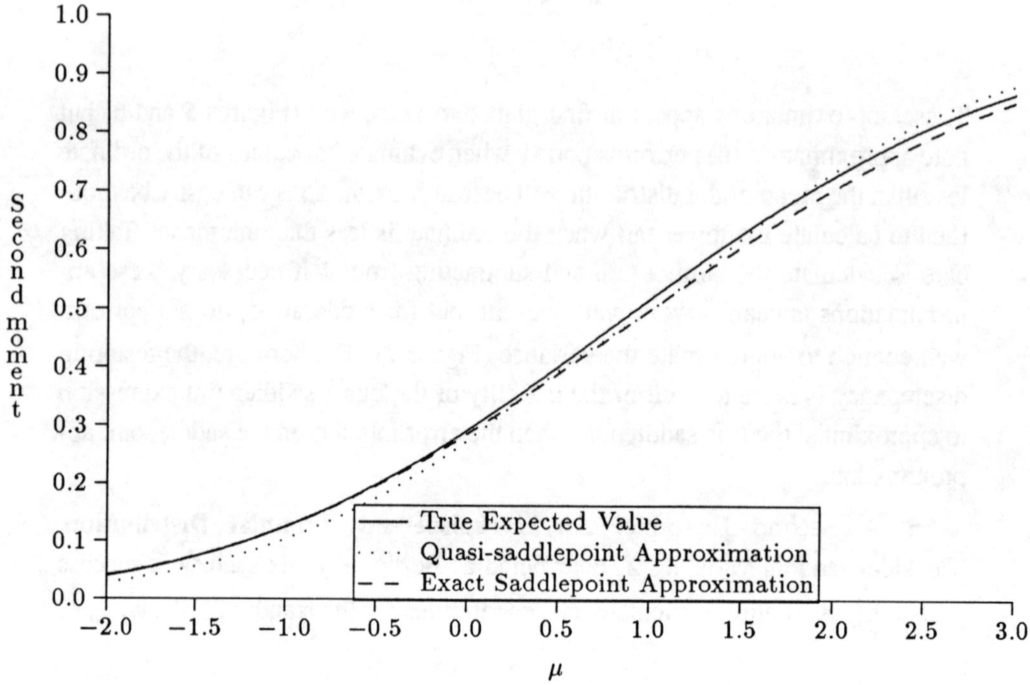

Fig. 6: Approximations to the Second Moment of the Resulting Parameter

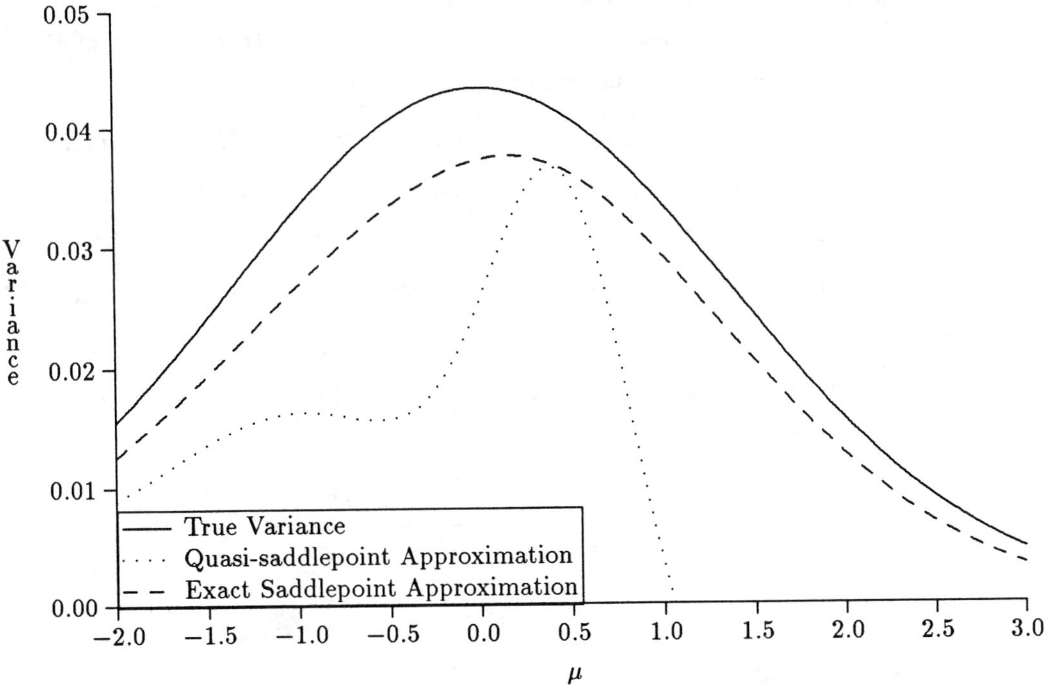

Fig. 7: Approximations to the Variance of the Resulting Parameter

the centered variables $X_i - p_i$. Were p_i all the same, exact tail probability calculations would be trivial. Were they distributed with some postulated distribution, techniques described for the previous example could be used to calculate unconditional success probabilities and also tail probabilities. Under more lax assumptions about their distribution, one might use a normal approximation to calculate unconditional tail probabilities. Unfortunately, the distribution of S_n has much lighter tails than the normal, and we might hope for better tail behavior. This section investigates what information about the p_i is necessary to apply the quasi-saddlepoint techniques to approximate tail probabilities.

Choose $p \in (0, 1)$. We will approximate the exact cumulant generating function of S_n, $K_n(t) = \Sigma_{i=1}^{n} [\log(p_i(\exp(t \ /\sqrt{n}) - 1) + 1) - p_i t/\sqrt{n}]$, by the cumulant generating function that arises when all of the p_i are set equal to p: $L_n(t) = n\log(p(\exp(t \ /\sqrt{n}) - 1) + 1 - \sqrt{n}pt$. Solving the saddlepoint equation for L_n, $L_n(t_a) = s$, yields $t_a = -\sqrt{n}\log\mathrm{it}(p + s/\sqrt{n}) - \log it(p)$]. Note that $p + s/\sqrt{n}$ is the average of indicator variables, and hence always in [0, 1]. The difference between K_n and L_n is given by

$$a_n(t) = \sum_{i=1}^{n} \log\left[1 - \frac{(p_i - p)(\exp(t \ \sqrt{n}) - 1)}{1 + p(\exp(t \ \sqrt{n}) - 1) - (p_i - p)t \ \sqrt{n}}\right].$$

Hence

$$a'_n(t_a) = \frac{1}{\sqrt{n}} \sum_{i=1}^{n} \frac{p_i - p}{[1 + p_i(\exp(t_a \ \sqrt{n}) - 1)][1 + p(\exp(t_a \ \sqrt{n}) - 1)]} - (p_i - p).$$

Lengthy but unenlightening calculations show that the third derivative of a_n is bounded over the possible choices of the p_i, and hence the error incurred by substituting L for K in (6) is of order $O(1/\sqrt{n})$ uniformly over choices of the p_i. This choice of L_n satisfies the requirements i-iv of section 1 regardless of the relation between the p_i and p. The quality of this approximation improves as p moves to represent a more central value for the p_i.

If we now assume that the p_i are distributed with mean p and some variance σ^2, the Central Limit Theorem implies that $a''_n(0)$ is asymptotically $\sigma^2 + O_p(1/\sqrt{n})$. Hence $a'_n(t_a)$ is asymptotically $t_a\sigma^2 + O_p(1/\sqrt{n})$. Also, $L''_n(t_a) = p(1-p)^2 + O(1/\sqrt{n})$. The resulting conditional tail probability approximation as a function of the random variables p_i is then

$$\exp[nL(t_a/\sqrt{n}) - t_a s + \frac{1}{2}(t_a)^2 L''(t_a\sqrt{n})]$$

$$[1 - \Phi\left(t_a\left(\sqrt{p}(1-p) - \frac{\sigma^2}{(1-p)\sqrt{p}}\right)\right)] + O_p(1/\sqrt{n}).$$

Hence this is also the unconditional probability to order $O(1/\sqrt{n})$.

Appendix 1. Order of the Error term in the Density Approximation. I aim to show that the error term in (3) is of the order stated. It suffices to prove the following theorem:

Theorem. Let $\{p_t(u)\}$ be a family of functions indexed by a parameter t, with a power series representation

$$p_t(u) = \sum_{k=3}^{m} n^{\frac{2-k}{2}} \frac{\beta_{t,k}}{k!} u^k + n^{\frac{1-m}{2}} r_{m,t}(u) \tag{9}$$

where the coefficients $\beta_{t,k}$ are bounded t in for all k, and a function $r_{m,t}(u)$ such that $|r_{m,t}(u)| \le C_m |u|^m$, then the inverse Fourier transform of $\exp(u^2/2 + p_t)$ evaluated at α is

$$\sum_{k=0}^{m-2} \sum_{j=0}^{\infty} n^{-\frac{k}{2}} \gamma_{jk} h_j(\alpha) + R_t(n), \tag{10}$$

where $\gamma = \Gamma(\beta)$ and

$$R_t(n) = o\left(n^{-\frac{m+1}{2}}\right)$$

uniformly in t. Dependence of γ is suppressed.

Proof: To see this, observe that

$$\exp(p(u)) = \left[\prod_{k=3}^{m} \exp\left[\frac{\beta_{t,k}}{k!} u^k n^{\frac{2-k}{2}}\right]\right] \exp\left[n^{\frac{1-m}{2}} r_{m,t}(u)\right]$$

$$= \left[\prod_{k=3}^{m} \sum_{r \le \frac{m}{k-2}} \frac{1}{r!}\left(\frac{\beta_{t,k}}{k!}\right) r n \frac{(2-k)r}{2} u^{rk} + R_{t,m,k}(u)\right]\left(1 + n^{\frac{1-m}{2}} R_{m,t}(u)\right) \tag{11}$$

where

$$R_{t,m,k}(u) = \frac{w_{k,m,n}^{\left[\frac{m}{k-2}\right]+1}}{\left(\left[\frac{m}{k-2}\right]+1\right)!}$$

for $w_{k,m,n}$ between 0 and $n^{\frac{2-k}{2}}\frac{\beta_{t,k}}{k!}u^k$, by Taylor's theorem.

Retaining all terms of order (2 - m)/2 and smaller in n in (11), and including all others in the remainder, gives

$$\exp(p(u)) = \sum_{k=0}^{m-2}\sum_{j=0}^{\infty} n^{-\frac{k}{2}}\gamma_{jk}u^j + R^*_{m,t}(u), \tag{12}$$

where $R^*_{m,t}(u)$ is of the form $u^m n^{\frac{1-m}{2}}$ times sums of products of the β's, negative powers of n, and positive powers of u, plus the same kind of expression times $r_{m,t}(u)$. It's absolute value can be bounded by $n^{\frac{1-m}{2}}$ times a polynomial $q_m(|u|)$ in $|u|$ whose coefficients are independent of n and t. Taking the inverse Fourier transform of (12) at α yields (10) plus a term bounded by $n^{\frac{1-m}{2}}$ times the normal density φ evaluated at α times a polynomial in α whose coefficients are independent of t and n.

Appendix 2. Order of the Error term in the Distribution Function Approximation I now calculate the order of the error $R_n^m(s)$ in (5). Denote the cumulative distribution function associated with $\sigma_a f_n(\mu_a+\sigma_a\xi;t_a)$ by D_n. Let

$$g(t_a) = \exp[nK(t_a/\sqrt{n}) - t_a\sqrt{n}K'(t_a/\sqrt{n})].$$

Then, using integration by parts,

$$|R_n^m(s)| = g(t_a)\left|\int_{-\alpha(t_a)}^{\infty}\exp[-z_a\xi]\,(f_n(\mu_a+\sigma_a\xi;t_a) - e_m(\xi;\beta^n))\,d\xi\right|$$

$$= g(t_a)\Big|\,(D_n(\xi) - E_m(\xi;\beta^n))\exp[z_a\alpha(t_a)] +$$

$$\int_{-\alpha(t_a)}^{\infty}\exp[-z_a\xi]\,z_a(D_n(\xi) - E_m(\xi;\beta^n))\,d\xi\Big|$$

$$\leq 2\left(\sup\left|\left(D_n(\xi) - E_m(\xi;\beta^n)\right)\right|\right)$$

$$\times \exp\left[nK(t_a/\sqrt{n}) - t_a\sqrt{n}K'(t_a/\sqrt{n}) + t_a\sqrt{n}a'(t_a/\sqrt{n})\right]$$

$$= 2\left(\sup\left|\left(D_n(\xi) - E_m(\xi;\beta^n)\right)\right|\right)\exp\left[nK(t_a/\sqrt{n}) - t_a\sqrt{n}L'(t_a/\sqrt{n})\right]$$

The Berry-Esseen theorem tells us that $\sup\left|(D_n(\xi) - E_m)(\xi;\beta^{n,t_a}))\right|$ is of order $o\left(n^{\frac{2-m}{2}}\right)$. Hence the error is always the order of the first term omitted, and our error bounds can be sharpened to $O\left(n^{\frac{1-m}{2}}\right)$. Furthermore, the Berry-Esseen bound involves the quantities β, which are uniformly bounded in t_a. By assumption,

$$\exp\left[nK(t_a/\sqrt{n}) - \sqrt{n}t_a L'(t_a/\sqrt{n})\right]$$

is uniformly bounded, and the error bounds given above hold uniformly in t_a.

References

[1] Blackwell, D., and Hodges, J.L. (1959). The probability in the extreme tail of a convolution. *Ann. Math. Statist.* **30** 1113-1120.

[2] Daniels, H.E. (1954). Saddlepoint Approximations in Statistics. *Ann. Math. Statist.* **25** 614-649.

[3] Daniels, H.E. (1987). Tail probability approximations. *Intern. Statist. Rev.* **55** 37-46.

[4] Feller, W. (1971). A*n Introduction to Probability Theory and its Applications* , II. Wiley.

[5] Lugannani, R. and Rice, S.O. (1980). Saddlepoint approximation for the distribution of the sum of independent random variables. *Adv. Appl. Prob.* **12** 475-490.

[6] McCullagh, P. (1987). *Tensor Methods in Statistics* , Chapman and Hall.

[7] Reid, N. (1988). Saddlepoint methods. *Statist. Sci.* **2** 213-238.

[8] Robinson, J. (1982). Saddlepoint approximations for permutation tests and confidence intervals. *J. Roy. Statist. Soc. Ser. B* **44** 91-101.

U-STATISTICS AND DOUBLE STABLE INTEGRALS

Joop Mijnheer
Leiden University and University of North Carolina

Summary

We derive the tail behaviour of the double stable integral

$$I(h) = \int_0^1\int_0^1 h(x, y)\, X(dx)\, X(dy),$$

where X is a completely asymmetric stable process.

1. Introduction. First we shall show the relation between the double stable integral and a simple U-statistic. Let $\{X(t): 0 \le t < \infty\}$ be a completely asymmetric stable process with characteristic exponent $\alpha \in (0,1)$ and $\beta = 1$. For the theory of stable distributions we refer to Gnedenko-Kolmogorov [GK 54], Breiman [Bre 68] or Feller[Fel 71]. A summary can be found in Mijnheer [Mij 75]. We use the notation as used in [Mij 75]. See [Mij 75] section 3.2 for a review of properties of stable processes. The random variables X_i, $i = 1, 2, \ldots$ are i.i.d. and have the same distribution as $X(1)$. $X \overset{d}{=} Y$ means that X and Y have the same distribution. $X \in D(\alpha, \beta)$ (resp. $D_N(\alpha,\beta)$) means that X belongs to the domain of (resp. normal) attraction of the stable distribution with parameters α and β. Then we have

$$n^{-2/\alpha} \sum_{i=1}^{n} \sum_{\substack{j=1 \\ i \ne j}}^{n} X_i X_j \overset{d}{=} n^{-2/\alpha} \sum_{i \ne j}\sum \{X(i) - X(i-1)\}\{X(j) - X(j-1)\}$$

$$\overset{d}{=} \sum_{i \ne j}\sum \{X(in^{-1}) - X((i-1)n^{-1})\}\{X(jn^{-1}) - X((j-1)n^{-1})\}.$$

This quadratic form is in a natural way related to the double stable integral

$$I(h) = \int_0^1\int_0^1 h(x, y)\, X(dx)\, X(dy) \tag{1.1}$$

where the function h is given by

$$h(x, y) = \begin{cases} 1 & 0 \le x, y \le 1 \ \ and \ \ x \ne y \\ 0 & \text{otherwise}. \end{cases} \tag{1.2}$$

The existence of double stable integrals is proved in Szulga and Woyczynski [SW 83]. Easy to investigate are integrals where the function h is of the type

$$h(x,y) = \phi(x)\phi(y). \tag{1.3}$$

For Wiener--Ito integrals we can restrict ourselves to functions h of this particular structure. See Denker [Den 85], lemma 2.2.3. In Section 3 we derive the tail behaviour of the integral (1.1) in the case h is defined by (1.2). In Section 4 we give the behaviour when h satisfies (1.3).

$\sum_{i=1}^{n}\sum_{\substack{j=1 \\ i\neq j}}^{n} X_i X_j$ is a (simple) example of a U-statistic. These U-statistics have been introduced by Hoeffding [Hoe 48]. For the general theory of U-statistics see Serfling [Ser 80], Shorack and Wellner [SW 86] and a review of Dehling [Deh 85].

Next we summarize the limit behaviour of the mentioned U-statistic. We distinguish several cases.

Case I. X_i has a finite second moment. $EX_1 = \mu$ and $\sigma^2(X_1) = \sigma^2$.

Case Ia. The **non-degenerate** case: $\mu \neq 0$. We write $X_i = \mu + \sigma U_i$, $i = 1,2, \ldots$ where U_i, $i = 1,2, \ldots$ are i.i.d. with $EU_i = 0$ and $\sigma^2(U_i) = 1$. Then we have

$$\begin{aligned} Y_n &= n^{-1}(n-1)^{-1}\sum_{i\neq j}\sum X_i X_j \\ &= \mu^2 + 2n^{-1}\mu\sigma\sum U_i + \sigma^2 n^{-1}(n-1)^{-1}\left\{(\sum U_i)^2 - \sum U_i^2\right\}. \end{aligned} \tag{1.4}$$

This implies

$$n^{1/2}(Y_n - \mu^2) = 2\mu\sigma n^{-1/2}(\sum U_i) + \sigma^2(n-1)^{-1}n^{-1/2}\left\{(\sum U_i)^2 - \sum U_i^2\right\}. \tag{1.5}$$

It follows from the central limit theorem that the first term on the right hand side of (1.5) has a normal limit distribution with expectation 0 and variance $(2\mu\sigma)^2$. The second term on the right converges in probability to 0 for $n \to \infty$, by the central limit theorem and the law of large numbers.

Case Ib. In the **degenerate** case, i.e. $\mu = 0$, we have

$$Y_n = \sigma^2 n^{-1}(n-1)^{-1}\left\{(\sum U_i)^2 - \sum U_i^2\right\}. \tag{1.6}$$

Thus

$$nY_n\sigma^{-2} = (n-1)^{-1} n\left\{(n^{-1/2}\sum U_i)^2 - n^{-1}\sum U_i^2\right\} \tag{1.7}$$

which converges by the central limit theorem and the law of large numbers to a random variable $\chi_1^2 - 1$, where χ_1^2 has a chi--square distribution with one degree of freedom.

Case II. Let $X_1 \in D(\alpha, \beta)$ with $1 < \alpha < 2$. Then $\mu = EX_1$ exists and we can write $X_i = \mu + \sigma U_i$, where $EU_i = 0$ and σ is some scale parameter. (Remark that the variance of X_1 is infinite.)

Case IIa. In the non--degenerate case we have (1.4). Now there exists a slowly varying function h such that $n^{-1/\alpha} h^{-1}(n) \sum_{i=1}^{n} U_i$ converges in distribution to a stable random variable with distribution function $F(\cdot; \alpha, \beta)$. We write

$$h^{-1}(n)\, n^{1-1/\alpha}(Y_n - \mu^2) = 2\mu\sigma h^{-1}(n)\, n^{-1/\alpha} \sum_{i=1}^{n} U_i$$

$$+ n^{-1/\alpha}(n-1)^{-1} h^{-1}(n)\,\sigma^2\left\{(\sum U_i)^2 - \sum U_i^2\right\} \tag{1.8}$$

Thus the first term on the right hand side has a stable limit distribution. Since $1 < \alpha < 2$ it follows that $n^{-1/\alpha}(n-1)^{-1} h^{-1}(n)\,(\sum U_i)^2$ converges to 0 in probability. The random variable $U_i^2 \in D(\alpha/2, 1)$; thus there exists a slowly varying function h_1 such that $n^{-2/\alpha} h_1^{-1}(n) \sum U_i^2$ converges in distribution to a stable random variable with distribution function $F(\cdot;\ \alpha/2, 1)$. Hence it follows that $n^{-1/\alpha}(n-1)^{-1} h^{-1}(n) \sum U_i^2$ converges in probability to 0.

General U-statistics in this case are considered in Malevich and Abdalimov [MA 77]. For many random variables $X_i \in D(\alpha, \beta)$ we can choose $h_1 = h^2$. But this is not in general true. From now on we restrict ourselves to those random variables in the domain of attraction where we have $h_1 = h^2$.

Case IIb. In the degenerate case we have

$$\sum_{i \neq j}\sum X_i X_j = \sigma^2 \left\{ (\sum U_i)^2 - \sum U_i^2 \right\}.$$

As in Case IIa and with the assumption made above, we have that the distribution of $n^{-1/\alpha} h^{-1}(n) \sum U_i$ converges weakly to $F(\cdot;\ \alpha,\beta)$ and the distribution of $n^{-2/\alpha} h^{-2}(n) \sum U_i^2$ to $F(\cdot;\ \alpha/2,1)$. Thus the statistics $\sigma^{-2} n^{-2/\alpha} h^{-2}(n) \sum_{i \neq j}\sum X_i X_j$ converge in distribution to the random variable $S_\alpha^2 - S_{\alpha/2}$, where S_α and $S_{\alpha/2}$ are dependent stable random variables with distribution functions $F(\cdot;\ \alpha,\beta)$ and $F(\cdot;\ \alpha/2,1)$.

This is a special case of a theorem on products of stable random variables. See Avram and Taqqu [AT 86].

We delete the case $\alpha = 1$ for well-known normalizing difficulties.

Case III. $X_1 \in D(\alpha, \beta)$ with $0 < \alpha < 1$. The random variable X_1 has no finite expectation. This case is the same as Case IIb.

2. The Characteristic Function. In this section we consider i.i.d. positive random variables $X_1, X_2, \ldots$ with common distribution function G given by

$$1 - G(x) = x^{-\alpha} \quad x > 1 \quad 0 < \alpha < 1. \tag{2.1}$$

Thus we have $X_1 \in D_N(\alpha, 1)$. Let $U_1, U_2, \ldots$ be i.i.d. uniformly distributed on $(0,1)$. Thus $G^{-1}(U_1) \stackrel{d}{=} X_1$. Then we have

$$\begin{aligned} T_n = \sum_{i \neq j}\sum X_i X_j &\stackrel{d}{=} \{\sum_{i=1}^{n} G^{-1}(U_i)\}^2 - \sum_{i=1}^{n} \{G^{-1}(U_i)\}^2 \\ &\stackrel{d}{=} \{\sum_{i=1}^{n} U_i^{-1/\alpha}\}^2 - \sum_{i=1}^{n} U_i^{-2/\alpha} \\ &= \{\sum_{i=1}^{n} U_{(i)}^{-1/\alpha}\}^2 - \sum_{i=1}^{n} U_{(i)}^{-2/\alpha}, \end{aligned}$$

where $U_{(1)} < , \cdots , < U_{(n)}$ a.s.

$$= 2U_{(1)}^{-1/\alpha}\{\sum_{i=2}^{n} U_{(i)}^{-1/\alpha}\} + \{\sum_{i=2}^{n} U_{(i)}^{-1/\alpha}\}^2 - \sum_{i=2}^{n} U_{(i)}^{-2/\alpha}.$$

Given $U_{(1)} = u$ we have

$$T_n| U_{(1)} = u \stackrel{d}{=} 2u^{-1/\alpha}\{\sum_{i=2}^{n} \overline{U}_i^{\,-1/\alpha}\} + \{\sum_{i=2}^{n} \overline{U}_i^{\,-1/\alpha}\}^2 - \sum_{i=2}^{n} \overline{U}_i^{\,-2/\alpha} \tag{2.2}$$

where $\overline{U}_2, ..., \overline{U}_n$ are i.i.d. and uniformly distributed on (u,1). $\overline{U}_2$ has a finite expectation and finite variance. We have for $u \downarrow 0$

$$\begin{aligned} \mu := \quad E\overline{U}_2^{\,-1/\alpha} &= (1-u)^{-1}\alpha(1-\alpha)^{-1}(u^{-(1-\alpha)/\alpha} - 1) \\ &\sim (1-u)^{-1}\alpha(1-\alpha)^{-1}u^{-1/\alpha+1} \end{aligned} \tag{2.3}$$

and

$$\begin{aligned} \sigma^2 := \quad E\overline{U}_2^{\,-2/\alpha} &= (1-u)^{-1}\alpha(2-\alpha)^{-1}(u^{-(2-\alpha)/\alpha} - 1) \\ &\sim (1-u)^{-1}\alpha(2-\alpha)^{-1}\upsilon^{-2/\alpha+1}. \end{aligned} \tag{2.4}$$

This implies that

$$\sigma^2(\overline{U}_2^{\,-1/\alpha}) \sim \sigma^2 \quad \text{for} \quad u\downarrow 0.$$

Given $U_{(1)} = u$, the random variable

$$V_{n-1} = \left(\sum_{i=2}^{n} \overline{U}_i^{\,-1/\alpha}\right)^2 - \sum_{i=2}^{n} \overline{U}_i^{\,-2/\alpha} \tag{2.5}$$

is again a U-statistic. Define, for $i = 1, ..., n$ $Y_i = \overline{U}_i^{-1/\alpha} - \mu$. Then we have

$$V_{n-1} = EV_{n-1} + \mu(n-2)\sum_{i=2}^{n} Y_i + \sum_{i=2}^{n}\sum_{\substack{j=2 \\ i\neq j}}^{n} Y_iY_j. \tag{2.6}$$

One easily computes for small u

$$E2n^{-2/\alpha}u^{-1/\alpha}\left(\sum_{i=2}^{n} \overline{U}_i^{-1/\alpha}\right) \approx 2\alpha(1-\alpha)^{-1}(nu)^{-2/\alpha+1},$$

$$En^{-2/\alpha}V_{n-1} \approx \alpha^2(1-\alpha)^{-2}(nu)^{-2/\alpha+2},$$

$$\sigma^2\left(2u^{-1/\alpha}n^{-2/\alpha}\left(\sum_{i=2}^{n}\overline{U_i}^{-1/\alpha}\right)\right)\approx 4\alpha(2-\alpha)^{-1}(nu)^{-4/\alpha+1},$$

$$\sigma^2\left(n^{-2/\alpha}\mu(n-2)\sum_{i=2}^{n}Y_i\right)\approx c(nu)^{-4/\alpha+3}, \text{ for some constant c}$$

and

$$\sigma^2\left(n^{-2/\alpha}\sum_{i=2}^{n}\sum_{\substack{j=2\\ i\neq j}}^{n} Y_iY_j\right)\approx (nu)^{-4/\alpha+2}.$$

We shall derive the characteristic function (cf.) f^* of the limit distribution of

$$n^{-2/\alpha}T_n^* = 2n^{-2/\alpha}U_{(1)}^{-1/\alpha}\sum_{i=2}^{n}\overline{U_i}^{-1/\alpha}+n^{-2/\alpha}EV_{n-1}. \tag{2.7}$$

In section 3 we shall give an estimate for the difference of this cf. f^* and the cf. f of the double stable integral for small values of the argument. For technical reasons we restrict ourselves to the case $\frac{1}{2}\leq\alpha<1$. See section 3.

One easily obtains that $n\,U_{(1)}$ converges for $n\to\infty$, in distribution to the exponential distribution. For $k_n\to\infty$ and $k_n=o(n)$ for $n\to\infty$ we have

$$P((nk_n)^{-1}<U_{(1)}<n^{-1}k_n)\to 1 \tag{2.8}$$

for $n\to\infty$.

Take $t>0$

$$Ee^{itn^{-2/\alpha}T_n^*} = E_{U_{(1)}}\left\{Ee^{itn^{-2/\alpha}T_n^*}\,\middle|\,U_{(1)}\right\}$$

$$= n\int_0^1(1-u)^{n-1}\left\{Ee^{itn^{-2/\alpha}T_n^*}\,\middle|\,U_{(1)}=u\right\}du.$$

From (2.7) we have

$$E\left(e^{itn^{-2/\alpha}T_n^*}\,\middle|\,U_{(1)}=u\right) = Ee^{2itn^{-2/\alpha}u^{-1/\alpha}\sum_{j=2}^{n}\overline{U_j}^{-1/\alpha}}e^{itn^{-2/\alpha}n(n-1)\mu^2}$$

$$= e^{itn^{-2/\alpha} n(n-1)\mu^2} \left\{ E e^{2itn^{-2/\alpha} \overline{U}_2^{-1/\alpha}} \right\}^{n-1},$$

where

$$E e^{2itn^{-2/\alpha} u^{-1/\alpha} \overline{U}_2^{-1/\alpha}} = (1-u) \int_u^1 e^{2itn^{-2/\alpha} u^{-1/\alpha} y^{-1/\alpha}} dy.$$

Consider the integral

$$I = (1-u)^{-1} \int_u^1 (e^{2itn^{-2/\alpha} u^{-1/\alpha} y^{-1/\alpha}} - 1)\, dy$$

$$= (1-u)^{-1} \alpha 2^{\alpha} t^{\alpha} n^{-2} u^{-1} \int_{2n^{-2/\alpha} u^{-1\alpha} t}^{2n^{-2/\alpha} u^{-2/\alpha} t} (e^{iv} - 1)\, v^{-1-\alpha} dv.$$

This integral is well-known in the derivation of the characteristic function of a stable random variable. We define

$$\phi(x) = \int_0^x (e^{iv} - 1)\, v^{-1-\alpha} dv. \tag{2.9}$$

We have

$$\phi(\infty) = -\alpha^{-1} \Gamma(1-\alpha)\, e^{-\pi i(\alpha/2)}. \tag{2.10}$$

See Laha and Rohatgi [LR79] p. 333. Thus

$$f^*(t) = \lim_{n\to\infty} E e^{itn^{-2/\alpha} T_n^*}$$

$$= \lim_{n\to\infty} n \int_{(nk_n)^{-1}}^{k_n n^{-1}} (1-u)^{n-1} (1+I)^{n-1} e^{it\alpha^2 (1-\alpha)^{-2} (un)^{-2/\alpha+2}} du$$

$$= \lim_{n\to\infty} \int_{k_n^{-1}}^{k_n} e^{-nu} e^{nI} e^{it\alpha^2(1-\alpha)^{-2}(un)-2/\alpha+2} dnu \tag{2.11}$$

$$= \int_0^\infty e^{-y} e^{it\alpha^2 (1-\alpha)^{-2} y^{-2/\alpha+2}} e^{\alpha 2^{\alpha} t^{\alpha} y^{-1} \phi(2ty^{-2/\alpha})} dy.$$

3. Tail Behaviour of the Stable Integral. In order to obtain the behaviour of the tail of the double stable integral we need the expansion of its cf. f near the origin. We rewrite the cf. f^* given in (2.11). Let

$$\phi_1(x) = \int_0^x (\cos v - 1) v^{-1-\alpha} dv \tag{3.1}$$

$$\phi_2(x) = \int_0^x \sin v \; v^{-1-\alpha} dv \tag{3.2}$$

and let the function g_t be defined by

$$g_t(y) = \begin{cases} e^{-y} e^{\alpha 2^\alpha t^\alpha y^{-1} \phi_1(2ty^{-2/\alpha})} & \text{for } 0 < y < \infty \\ 0 & \text{else} \end{cases} \tag{3.3}$$

Then $\phi_1(\infty)$ and $\phi_2(\infty)$ follow from (2.10). We have

$$f^*(t) = \int_0^\infty e^{it\alpha^2(1-\alpha)^{-2} y^{-2/(\alpha+2)}} e^{i\alpha 2^\alpha t^\alpha y^{-1} \phi_2(2ty^{-2/\alpha})} g_t(y)\, dy. \tag{3.4}$$

Since for $y > (2t)^{\alpha/2}$ both $2^\alpha t^\alpha y^{-1}$ and $2ty^{-2/\alpha}$ are small, we easily obtain

$$\int_{(2t)^{\alpha/2}}^\infty g_t(y)\, dy = \int_{(2t)^{\alpha/2}}^\infty e^{-y} dy + O(t^\alpha) \tag{3.5}$$

for $t \downarrow 0$. Obviously we have

$$\int_0^{\alpha 2^\alpha t^\alpha} g_t(y)\, dy = O(t^\alpha) \tag{3.6}$$

for t 0. Then

$$\int_{\alpha 2^\alpha t^\alpha}^{(2t)^{\alpha/2}} g_t(y)\, dy = \int_{\alpha 2^\alpha t^\alpha}^{(2t)^{\alpha/2}} \left\{ 1 - y + \alpha 2^\alpha t^\alpha y^{-1} \phi_1(2ty^{12/\alpha}) \right\} dy + \text{error}. \tag{3.7}$$

Using the definition of ϕ_1 and by partial integration we obtain

$$\alpha \int_{\alpha 2^\alpha t^\alpha}^{(2t)^{\alpha/2}} y^{-1} \phi_1(2ty^{-2/\alpha})\, dy = 2 \int_1^{\alpha^{-2/\alpha} 2^{-1} t^{-1}} z^{-1} \phi_1(z)\, dz$$

$$= 2\phi_1(\alpha^{-2/\alpha} 2^{-1} t^{-1}) \log(\alpha^{-2/\alpha} 2^{-1} t^{-1}) + O(t^\alpha) \quad \text{for } t \downarrow 0.$$

With similar calculations as above we show that the error in the right hand side of (3.7) is $O(t^\alpha)$ for $t\downarrow 0$. Combining the results in (3.5), (3.6) and the result above, we obtain

$$\int_0^\infty g_t(y)\,dy = 1+2^{1+\alpha}t^\alpha\phi_1(\alpha^{-2/\alpha}2^{-1}t^{-1})\log(t^{-1}) + O(t^\alpha)$$

for $t\downarrow 0$.

For the expansion of f^* we have to distinguish two cases: $\frac{1}{2}\le\alpha<1$ and $0<\alpha<\frac{1}{2}$. As mentioned before, we only consider the case $\frac{1}{2}\le\alpha<1$. See after (3.11). Now we have $t^{\alpha/(2(1-\alpha))} < t^\alpha < t^{\alpha/2}$.

$$f^*(t) = \int_0^\infty e^{it\alpha^2(1-\alpha)^{-2}y^{-2/\alpha+2}}\, e^{i\alpha 2^\alpha t^\alpha y^{-1}\phi_2(2ty^{-2/\alpha})}\, g_t(y)\,dy$$

$$= \int_0^{\alpha 2^\alpha t^\alpha} + \int_{\alpha 2^\alpha t^\alpha}^{(2t)^{\alpha/2}} + \int_{(2t)^{\alpha/2}}^\infty = I_1+I_2+I_3.$$

Obviously we have

$$I_1 = O(t^\alpha) \quad \text{for } t\downarrow 0,$$

For $y>(2t)^{\alpha/2}$ we have that $ty^{-2/\alpha+2}$, $2^\alpha t^\alpha y^{-1}$ and $2ty^{-2/\alpha}$ are small. Expansion of the integrand gives

$$I_3 = \int_{(2t)^{\alpha/2}}^\infty e^{-y}dy + O(t^\alpha) \quad \text{for } t\downarrow 0.$$

I_2 is the most interesting part. We have

$$I_2 = \int_{\alpha 2^\alpha t^\alpha}^{(2t)^{\alpha/2}} \left\{1+it\alpha^2(1-\alpha)^{-2}y^{-2/\alpha+2} + i\alpha 2^\alpha t^\alpha y^{-1}\phi_2(2ty^{-2/\alpha})\right\} g_t(y)\,dy \tag{3.9}$$

$$+ \text{ error.}$$

For $\alpha\ge\frac{1}{2}$ we have

$$t\int_{\alpha 2^\alpha t^\alpha}^{(2t)^{\alpha/2}} y^{-2/\alpha+2}dy = O(t^\alpha).$$

And, as in computation of the integral of g_t, we have

$$\alpha t^{\alpha} \int_{\alpha 2^{\alpha} t^{\alpha}}^{(2t)^{\alpha/2}} y^{-1} \phi_2 (2ty^{-2/\alpha})\, dy = 2t^{\alpha} \phi_2 (\alpha^{-2/\alpha} 2^{-1} t^{-1}) \log (\alpha^{-2/\alpha} 2^{-1} t^{-1}) + O(t^{2\alpha}).$$

One easily shows that the error on the right hand side of (3.9) is also $O(t^{\alpha})$ for $t \downarrow 0$. Combining the foregoing results, one obtains in the case $\frac{1}{2} \le \alpha < 1$

$$f^{*}(t) - 1 = ct^{\alpha} \log (t^{-1}) + O(t^{\alpha}) \tag{3.10}$$

for $t \downarrow 0$. We have

$$f(t) - f^{*}(t) = \lim_{n \to \infty} E_{U_{(1)}} \left[E\left\{ e^{itn^{-2/\alpha} T_n} \middle| U_{(1)} \right\} - E\left\{ e^{itn^{-2/\alpha} T_n^{*}} \middle| U_{(1)} \right\} \right]$$

$$= \lim_{n \to \infty} E_{U_{(1)}} \left[E e^{2itn^{-2/\alpha} U_{(1)}^{-1/\alpha} \sum_{j=2}^{n} \overline{U}_j^{-1/\alpha}} \left\{ e^{itn^{-2/\alpha} V_{n-1}} - e^{itn^{-2/\alpha} EV_{n-1}} \right\} \middle| U_{(1)} \right].$$

Applying the inequality

$$|e^{ix} - e^{iy}| \le |x - y|$$

and similar calculations as above, we obtain, for $\frac{1}{2} \le \alpha < 1$,

$$|f(t) - f^{*}(t)| = O(t^{\alpha}) \quad \text{for } t \downarrow 0.$$

This implies the following expansion for f

$$f(t) - 1 \sim ct^{\alpha} \log (1/t) \tag{3.11}$$

for $t \downarrow 0$.

In the case $0 < \alpha < 1/2$ it is more delicate to obtain an estimate for $f - f^{*}$. For that reason we delete this case.

From the theory of characteristic functions we obtain from the expansion of f

$$P(I(h) > x) \sim ax^{-\alpha} \log x \quad \text{as } x \to \infty.$$

See, for example, Feller [Fel. 71] section xvii.12, problem 14.

4. Extension. In the previous section we considered the integral

$$I(h) = \iint h(x,y) X(dx) X(dy)$$

where X is a completely asymmetric stable process and h is given by (1.2). We can extend the integral to functions given by (1.3).

THEOREM 4.1. Let $I(h)$ be given as above and

$$h(x,y) = \begin{cases} \phi(x)\phi(y) & for\ x \neq y \\ 0 & for\ x = y \end{cases}.$$

Suppose ϕ positive and $\int_0^1 \phi(x)^\alpha dx < \infty$. Then

$$I(h) \overset{d}{=} \left\{\int_0^1 \phi^\alpha(x)\,dx\right\}^{2/\alpha} \left\{S_\alpha^2 - S_{\alpha/2}\right\}$$

where S_α and $S_{\alpha/2}$ are dependent stable random variables.

PROOF.

$$I(h) = \lim\left[\{\sum_{i=1}^{n} \phi(i/n) n^{-1/\alpha} X_i\}^2 - \sum_{i=1}^{n} \phi^2(i/n) n^{-2/\alpha} X_i^2\right]$$

$$\overset{d}{=} \left[\left\{\int_0^1 \phi^\alpha(x)\,dx\right\}^{1/\alpha} X\right]^2 - \left\{\int_0^1 \phi^\alpha(x)\,dx\right\}^{2/\alpha} S_{\alpha/2}$$

$$= \left\{\int_0^1\int_0^1 h^\alpha(x,y)\,dxdy\right\}^{1/\alpha} \left\{S_\alpha^2 - S_{\alpha/2}\right\}.$$

See also section 1, case IIb. The first limit follows from properties of stable random variables. One shows the second limit by using characteristic functions. □

Remark. We can obtain the tail behaviour as in the case h satisfies (1.2). Note that the dependency of S_a and $S_{\alpha/2}$ depends on ϕ.

In Samorodnitsky and Szulga [SS 88] the asymptotic behaviour of the tail of $I(h)$ is given in the case of a <u>symmetric</u> stable process. They obtain if h satisfies the conditions of Theorem 4.1

$$P(|I(h)| > x) \sim C_\alpha(h) x^{-\alpha} \log x \quad \text{as } x \to \infty.$$

5. The Behaviour of the Extremal Term. In this section we consider the term $U_{(1)}^{-1/\alpha}U_{(2)}^{-1/\alpha}$ where $U_{(1)}$ and $U_{(2)}$ are the order statistics of a uniform distribution. Let Γ_i, $i = 1,2, \dots$ be the arrivals of a Poisson process. We have

$$
\begin{aligned}
P(U_{(1)}^{-1/\alpha}U_{(2)}^{-1/\alpha} > n^{2/\alpha}x) &= P(U_{(1)}U_{(2)} \le n^{-2}x^{-\alpha}) \\
&= P\left(\frac{\Gamma_1^{-1/\alpha}\Gamma_2^{-1/\alpha}}{\Gamma_n^{-2/\alpha}} > n^{2/\alpha}x\right) \\
\sim P(\Gamma_1^{-1/\alpha}\Gamma_2^{-1/\alpha} > x) &= P(\Gamma_1\Gamma_2 < x^{-\alpha}) \\
&= x^{-\alpha}\int_{x^{-\alpha/2}}^{\infty} y^{-1}e^{-y}dy + \int_0^{x^{-\alpha/2}} ye^{-y}dy \\
&= x^{-\alpha}\int_{x^{-\alpha/2}}^{\infty} y^{-1}e^{-y}dy + 1 - e^{-x^{-\alpha/2}} - x^{-\alpha/2}e^{-x^{-\alpha/2}} \\
&\sim \frac{1}{2}\alpha x^{-\alpha}\log x \quad \text{for large } x.
\end{aligned}
$$

We have seen a similar tail behaviour for the double α-stable integral.

References

[AT 86] F. Avram and M. S. Taqqu. Symmetric Polynomials of Random Variables Attracted to an Infinitely Divisible Law. Probab. Th. Rel. Fields **71**, 491-500 (1986).

[Bre 86] L. Breiman. Probability. Addison-Wesley, Reading (Mass.).

[Den 85] M. Denker. Asymptotic Distribution Theory in Nonparametric Statistics. F. Vieweg and Sohn, Braunschweig/Wiesbaden.

[Deh 85] H. Dehling. Almost sure approximations for U-statistics. Dependence in Probability and Statistics. E. Eberlein and M. S. Taqqu, eds. Progress in Prob. and Stat. Vol. **11**. Birkhauser, Boston. 129-136.

[Fel 71] W. Feller. An introduction to probability theory and its applications. Vol. II. Second Edition. Wiley, New York.

[GK 54] B. V. Gnedenko and A. N. Kolmogorov. Limit distributions for sums of independent random variables. Addison-Wesley, Reading (Mass.).

[Hoe] W. Hoeffding. A class of statistics with asymptotically normal distribution. Ann. Math. Statistics **19**, 293-325 (1948).

[LR 79] R. G. Laha and V. K. Rohatgi. Probability theory. Wiley, New York.

[MA 77] T. L. Malevich and B. Abdalimov. Stable limit distributions for U-statistics. Th. of Prob. and Appl. **22**, 370-377.

[Mij 75] J. L. Mijnheer. Sample path properties of stable processes. Math. Centre, Amsterdam.

[SS 88] G. Samorodnitsky and J. Szulga. An asymptotic evaluation of the tail of a multiple symmetric α-stable integral. Annals of Prob., Vol. **17**, 1503-1520 (1989).

[ST 88] G. Samorodnitsky and M. S. Taqqu. Multiple Stable Integrals of Banach-Valued Functions. Preprint.

[Ser 80] R. J. Serfling. Approximation Theorems of Mathematical Statistics. Wiley and Sons, New York.

[SW 86] G. R. Shorack and J. A. Wellner. Empirical processes with applications to statistics. Wiley and Sons, New York.

[SW 83] J. Szulga and W. A. Woyczynski. Existence of a Double Random Integral with Respect to Stable Measures. J. Mult. Anal. **13**, 194-201.

GENE CONVERSION AND THE INFINITE-SITES MODEL

Stanley A. Sawyer
Washington University

Abstract

The infinite-sites model assumes that there have been only simple point mutations since the common ancestor of a sample of aligned DNA sequences, and in particular no recombination or gene conversion. A statistical test for detecting a history of gene conversion within a sample of aligned DNA sequences is applied to various data sets. The results show strong evidence for multiple intragenic conversion events at two loci in the bacterium *E. coli* and in a repeated family in maize. The data suggest that the rate of these short conversion events may be much larger than the neutral mutation rate per base. Thus the basic assumptions of the infinite-sites model may often be violated in natural data sets. The same test applied to a gene in an RNA virus and to a sample of mitochondrial DNA did not show gene conversion.

Introduction. The basic genetic material or DNA of plants and animals can be thought of as a set of strings of letters from an alphabet {T, C, A, G}, with one string for each chromosome. Thus to a geneticist a mouse is the same as a tomato, since both have about the same amount of DNA. A gene or genetic locus is a segment of a chromosome that affects some trait. Experiments beginning in the middle 1960's showed that many enzymes were unexpectedly polymorphic; i.e., their genetic loci had genes of many different types. A protracted and sometimes heated debate arose about the selective significance of these polymorphisms. Some biologists felt that most of these observed enzyme polymorphisms were selectively neutral or nearly neutral, while others asserted that different enzymes were unlikely to be selectively equivalent, and so various forms of balancing selection must be involved (Kimura, 1968; Wills, 1973; see also Hartl, 1989; Hartl and Sawyer, 1991). In an attempt to resolve this controversy, Ewens (1972) derived a formula for the expected distribution of genes among allelic classes in a random sample from a population that is selectively neutral at that genetic locus. The infinite-alleles model in all of its various forms can be considered an extension of Ewens' (1972) work. The vast number of recent papers on the infinite-alleles model by Donnelly, Ethier, Griffiths, Hoppe, Hudson, Kaplan, Kurtz, Tavare, Watterson, and others shows the health and vigor of current mathematical research in this field.

The infinite-alleles model assumes an equilibrium selectively-neutral random-mating population subject to mutation, where each new mutant gene is of a type that is entirely new to the population. In 1972, most data on enzyme polymorphisms came from protein electrophoresis, which allows genes to be distinguished but gives no further information about them. Since 1983, DNA sequences of genes at polymorphic loci have become available. Foreshadowing this experimental advance, Watterson (1975) considered the infinite-alleles model where the alleles are DNA sequences with the assumption that each new mutation changes a single base (i.e., one of the letters T,C,A,G) at a previously undisturbed site. This model is now known as the infinite-sites model. An important feature of the infinite-sites model is that the pedigree of a sample can be read off from the DNA sequences if the sequence of the common ancestor is known (Felsenstein 1982, p380). One can often use "outgroups" to find information about the DNA sequence of the common ancestor. For example, if DNA sequences are known for seven squirrels and a hedgehog at a particular genetic locus, and if the squirrel sequences are polymorphic at a site for which one of the bases T,C,A,G also occurs in the hedgehog at that site, then the base in the hedgehog is probably in the common ancestor of both species. However, the assumption that all mutations since the common ancestor of the sample were simple point mutations at distinct sites is crucial. One of the purposes of this article is to show that this assumption may often be violated.

One way in which a multiple-base mutation can occur is if a recombination or crossover event occurs within a gene. This produces a gene composed of parts of the two parental genes, and is likely to be new to the population if the parental genes were distinct or if the partial genes were misaligned. Humans have about one crossover event per chromosome pair for each offspring, and the chromosomes are on the average about 10^8 bp (base pairs) long. The point mutation rate in higher animals is about 5×10^{-9} per site per year (Li *et al.* 1985), or about 10^{-7} per generation per site in humans. Thus perhaps about 10% of mutations in humans are due to crossover events within genes. However, the distribution of crossover points is highly nonuniform, and the ratio for particular genes may be either much less or much more than 10% .

A more important type of non-simple-point mutation is *gene conversion* within a gene. Gene conversion is any process that copies the bases from one segment of DNA to another segment of the same length, and can occur by a variety of mechanisms. The source and target DNA segments can be on the same chromosome, or else they can be on different chromosomes in the same individual or in different individuals. Table 1 illustrates a gene conversion event where the source and target DNA segments are 15bp long. Note that only the 4 bases that are flagged actually change in the target sequence. In observed data, this event would appear as a cluster of four point mutations. Since the new bases at these sites are the same as in the source sequence, the mutations are not independent. If the source and target segment happened to be identical, no change would occur. Gene mutations that are definitely due to gene conversion have been identified in humans (Slightom *et al.* 1980), mice (Weiss *et al.* 1983), and in many other creatures as well. Nagylaki (1988), Ohta (1986), Watterson (1989), and Griffiths and Watterson (1990) are a partial list of recent theoretical papers on gene conversion.

Table 1: An Illustration of Gene Conversion

······ctttaagcgcat GCAGAAGGCTTTAACcgaatgatcggt······
↓ ↓ ↓ ↓
······attgtgtttatt GCACATGGATCTAAC tagtgccggtga······

Many mechanisms for gene conversion depend on a combination of spontaneous DNA hybridization and mismatch repair. For example, the two DNA strands in each of the two DNA segments in Table 1 may accidentally separate, and then one strand from each of the two segments may temporarily bind together (or "hybridize") by the same chemical forces that normally bind the two strands in a single DNA segment. Enzymes that normally correct mismatches between the two strands in normal DNA may then force the second strand to be identical with the first. In general, the mismatch repairs may not all be in the same direction, and not all mismatches may be repaired, but Table 1 may occur. If the two mismatched strands then separate and re-attach with their original strands, and if the mismatch repair enzymes copy the introduced changes to the other DNA strand in the target DNA segment, then the changes illustrated in Table 1 will result.

Genes usually have an "upstream" control region followed by one or more "coding regions" and perhaps a trailing "downstream" control region. Triples of bases in coding regions form *codons* that specify unique amino acids. Proteins and enzymes are composed of chains of amino acids. Since there are 64 possible codons and only 20 amino acids, the mapping cannot be one-to-one, with non-uniqueness typically in the third position of codons. Table 2 gives the first 36 bases (and the resulting amino acids) in the coding region of a typical gene for the enzyme 6-phosphogluconate dehydrogenase (abbreviated "*gnd*") in the bacterium *Escherichia coli.* Note that the amino acid glycine is represented in Table 2 by two different codons (GGC and GGT), and phenylalanine by two codons TTT and TTC. Since TTC and TTT are the only codons for phenylalanine, the third positions of these codons are called *two-fold degenerate* or *two-fold silent sites* . Since all four codons GGx code for glycine, the third positions of those codons are called *four-fold degenerate* or *four-fold silent*. There is only one three-fold degenerate amino acid, isoleucine (Ile). We implicitly assume that silent changes in DNA are selectively neutral, since the same protein is encoded. Although selection can act in various ways on silent DNA differences, the average effect of selection on configurations at silent sites appears to be small (Sawyer *et al.* 1987; Hartl and Sawyer, 1990).

Table 2: The Initial Codons and Amino Acids in *gnd*

GCA	GAA	GGC	TTT	AAC	TTC	ATT	GGT	ACC	GGT	GTT	TCC. .
Ala	Glu	Gly	Phe	Asn	Phe	Ile	Gly	Thr	Gly	Val	Ser. . .

We next describe a statistical test for a pattern of gene conversion in a sample of DNA sequences that looks for nonrandom pairwise concordances at silent sites. We then describe some data sets for which this test is extremely significant for apparent within-locus gene conversion.

Statistical Tests for Gene Conversion. Given an aligned sample of DNA sequences, a gene conversion event will tend to produce a segment in which two sequences are identical. If the segment is long, and if there has been no subsequent mutations in the segment, the set of polymorphic sites that distinguish the two sequences may have a highly nonuniform distribution. Under various assumptions one can assign a conservative P-value for this apparent gene conversion by classical probabilistic arguments (Stephens 1985). However, if the segments undergoing gene conversion are short, or if they once were long but have since been altered by mutation, then individual gene conversion events will not be detectable by this approach. The procedure we describe below is a permutation test that is designed to detect the result of gene conversion, while not necessarily being able to detect specific occurrences (Sawyer 1989).

In general, permutation tests have three basic components. The first is a null hypothesis H_0, in common with all statistical tests. In our case, this is the assumption that there has been no recombination or gene conversion since the common ancestor of the sample. The distribution of bases at silent polymorphic sites would then be determined by independent neutral mutation acting on the same pedigree at all sites. The test we consider below will depend only on the silent polymorphic sites and their relative order along the sequenced region. It is useful to think of this data as an $n \times s$ matrix of bases T, C A, G, where n is the number of aligned sequences and s is the number of silent polymorphic sites. Given H_0, the columns in this matrix have a statistical distribution depending only on the ancestral base and the degeneracy k at a particular site. If we relabel the bases at each site as 1, ..., k where the base relabeled as i forms the ith most common nucleotide at that site, the columns at silent sites will have a distribution given H_0 depending only on the degeneracy k.

The second component of permutation tests is a class of permutations G of the data preserving H_0. More precisely, we assume that the data are the result of observations of random variables whose joint distribution given H_0 is not changed by the permutations in G. The class G in our case will consist of all permutations of the columns of the $n \times s$ matrix that permute silent polymorphic sites of the same degeneracy among themselves. (We treat silent polymorphic sites of irregular degeneracy as a single class for simplicity.) If H_0 is correct, the observed data set should appear as a typical member of the class of permuted data sets corresponding to the permutations in G. If the observed data set is atypical, we would suspect hidden structure inconsistent with H_0. The class G is intended to break up a pattern of short pairwise-conserved segments that may not be immediately apparent.

The third component of a permutation test is a "score"; i.e., a statistic applied to possible data sets that should be small if H_0 is correct, but large if H_0 fails in some way that we wish to detect. The P-value of the permutation test is defined as the probability that the data set generated by a randomly chosen permutation from G will have a score larger than or equal to the observed score. Since G will usually be too large for the P-value to be calculated exactly, we will approximate the P-value by using 10,000 randomly chosen permutations from G.

The score we choose is as follows. If two of the n sequences are compared, they will differ in a set of $d \leq s$ silent polymorphic sites. These d sites partition the array of silent polymorphic sites into d + 1 segments, which we will call the *condensed fragments* determined by this partition ("condensed" because we have thrown out the non-silent-polymorphic sites). Each condensed fragment corresponds to a segment from the sequenced region in which the two sequences are identical, except within amino acid polymorphic codon positions, and which is bounded either by two of the d discordant silent polymorphic sites or by one of the discordant sites and one end of the sequenced region. If x_i is the length of the ith condensed fragment, then $\Sigma x_i = s - d$. Each of the $n(n-1)/2$ pairs of sequences defines a set of $d_k + 1$ condensed fragments, where $1 \leq k \leq n(n-1)/2$. The score we will use is the sum of the squares of the condensed fragment lengths (abbreviated SSCF), which is defined as the double sum $\Sigma\Sigma\, x_i^2$ with summation over the $n(n - 1)/2$ pairs of sequences in the outer sum, and over the $d_k + 1$ condensed fragments determined by the kth sequence pair in the inner sum. Thus if $d_k = d$ for all k, the sum defining SSCF would have $(d + 1)n(n - 1)/2$ terms. The score SSCF is one of four similar scores defined in Sawyer (1989).

Applications of the SSCF Test. Results for a number of DNA sequence data sets are shown in Table 3. The first three numerical columns in Table 3 are the number n of aligned sequences, the number s of silent polymorphic sites, and the length of the sequenced region in bases. The first two rows are for two enzymes in *Escherichia coli* (*phoA* is an abbreviation for alkaline phosphatase). The observed score is highly significant for both *E. coli* genetic loci. Table 3 also gives the distance between the observed score and the mean of the permuted scores in terms of the standard deviation of the permuted scores. While this distribution is not normal, the fact that the observed value of SSCF for *gnd* is 7.21 standard deviations above the mean is consistent with the fact that the largest of 10,000 simulated values of SSCF was still well below the observed value. An analysis of the distribution of uncondensed fragment lengths shows a significant excess of fragments of length 70-200bp for *gnd* ($P < 10^{-3}$), and of 190bp and longer for *phoA* ($P \approx 0.03$). While SSCF was highly significant for both *gnd* and *phoA*, only the largest uncondensed fragment was significant by itself for *gnd*, and no fragments were significant for *phoA*, using a related test that applies to single fragments (Sawyer 1989). These observations suggest a pervasive pattern of intragenic recombination within both the *gnd* and *phoA* loci (see also DuBose *et al*, 1988, for *phoA*). None of the six largest uncondensed fragments in *gnd* were bounded by an endpoint of the sequenced region, which suggests a pattern of either short gene conversion events or else large gene conversion events punctuated by later mutation.

The test was nonsignificant when applied to a simulated *gnd* data set with mutation only. When 80 simulated gene conversion events for segments of 50-200bp were added to the simulated *gnd* data set, the results approximated those of the observed *gnd* data set (Table 3). This corresponds to a rate of gene conversion about 20 times that of point mutation per base. The SSCF test was also nonsignificant when applied to 13 strains of human influenza A virus. The human influenza data was not expected to show gene conversion because (1) the rapid mutation rates of RNA retroviruses such as influenza may

Table 3: SSCF Test Results for Gene Conversion

Data Set	Number of Seqs.	Number of Polys.	Number of Bases	P-value	(Obs.-Mean)[a] / S. D.
E. coli (*gnd* locus[b])	7	81	768	0	7.21
E. coli (*phoA* locus[c])	8	61	1413	0	5.33
Sim. *gnd*, mut. only[d]	7	78	768	0.2719	0.54
Sim. *gnd*, mutation and gene conversion[d]	7	83	768	0	7.46
Influenza type A[e]	13	52	678	0.6295	-0.38
Cytochrome c[f]	9	52	*82*	0.0966	1.36
Human globins[g]	3	36	1592	0	12.58
Maize heterochromatin[h]	11	36	180	0.0003	4.71
Primate mitochondria[i]	5	283	896	0.9625	-1.61

[a]Mean and S.D. are the mean and standard deviation of the scores for 10,000 random permutations of the data.

[b]Sawyer *et al.* (1987) [c]DuBose *et al.* (1988) [d]Sawyer (1989) [e]Buonagurio et al. *(*1986*)* [f]Sneath *et al.* (1975) [g]Slightom *et al.* (1980) [h]Dennis and Peacock (1984) [i]Brown *et al.* (1982)

obliterate any evidence for gene conversion, and (2) the virus strains were gathered at different times over a 50 year period. All polymorphic sites can be permuted (and used as potential discordant sites) instead of silent polymorphic sites, with similar results in most cases. However, when this test is used with all polymorphic sites, a significant result may be due to parallel selection or some other selective effect rather than gene conversion. When applied to nine bacterial protein sequences of 82 amino acid bases each, the result approached significance ($P \approx 0.097$; see Table 3). Since these bacteria are thought to be too distantly related for gene conversion to occur, the near-significance is most likely due to parallel evolution in portions of the sequenced region. The human globin entry in Table 3 relates to a single apparent gene conversion between two loci on the same human chromosome.

Dennis and Peacock (1984) have DNA sequences for 11 repeats of a 180bp segment in knob heterochromatin, which is a family with up to 10^6 tandem repeats in maize. Since these repeats are adjacent and similar, between-repeat gene conversion is expected by spontaneous hybridization. The SSCF test behaves as expected (Table 3). Gene conversion between mitochondria in different animals is thought to be quite rare. Consistent with this popular wisdom, the SSCF test detects no gene conversion in a sample of mitochondrial DNA sequences from seven primates (Table 3). See Sawyer (1989) and Hartl and Sawyer (1990) for more examples and for more biological detail about the examples in Table 3.

References

[1] Brown, W., E. Prager, A. Wang, and A. Wilson (1982). Mitochondrial DNA sequences of primates: tempo and mode of evolution. *J. Mol. Evol.* **18**, 225-239.

[2] Buonagurio, D., S. Nakada, J. Parvin, M. Krystal, P. Palese, and W. Fitch (1986). Evolution of human influenza A viruses over 50 years: rapid, uniform rate of change in NS gene. *Science* **232**, 980-982.

[3] Dennis, E. and W. Peacock (1984). Knob heterochromatin homology in maize and its relatives. *J. Mol. Evol.* **20**, 341-350.

[4] DuBose, R., D. Dykhuizen, and D. Hartl (1988). Genetic exchange among natural isolates of bacteria: recombination within *the phoA* locus of *Escherichia coli. Proc. Nat. Acad. Sci. USA* **85**, 7036-7040.

[5] Ewens, W. (1972). The sampling theory of selectively neutral alleles. *Theoret. Population Biol.* **3,** 87-112.

[6] Felsenstein, J. (1982). Numerical methods for inferring evolutionary trees. *Quarterly Review of Biology* **57**, p379-404.

[7] Griffiths, R. and G. Watterson (1990). The number of alleles in multigene families. To appear in *Theoret. Population Biol.*, **37**, p 110-123.

[8] Hartl, D. (1989). Evolving theories of enzyme evolution. *Genetics* **122**, 1-6.

[9] Hartl, D. and S. Sawyer (1991). Inference of selection and recombination from nucleotide sequence data. In press, *Jour. Evol. Biol.*

[10] Kimura, M. (1968). Evolutionary rate at the molecular level. *Nature* **217**, 624-626.

[11] Li, W.-H., C.-I. Wu, and C.-C. Luo (1985). A new method of estimating synonymous and nonsynonymous rates of nucleotide substitution considering the relative likelihood of nucleotide and codon changes. *Mol. Biol. Evol.* **2**, 150-174.

[12] Nagylaki, T. (1988). Gene conversion, linkage, and the evolution of multigene families. *Genetics* **120**, 291-301.

[13] Ohta, T. (1986). Actual number of alleles contained in a multigene family. *Genet. Res.* **48**, 119-123.

[14] Sawyer, S., D. Dykhuizen, and D. Hartl (1987). Confidence interval for the number of selectively neutral amino acid polymorphisms. *Proc. Nat. Acad. Sci. USA* **84**, 6225-6228.

[15] Sawyer, S. (1989). Statistical tests for detecting gene conversion. *Mol. Biol. Evol.* **6**, 526-538.

[16] Slightom, J., A. Blechl, and O. Smithies (1980). Human fetal G_γ and A_γ globin genes: complete nucleotide sequences suggest that DNA can be exchanged between these duplicated genes. *Cell* **21**, 627-638.

[17] Sneath, P., M. Sackin, and R. Ambler (1975). Detecting evolutionary incompatibilities from protein sequences. *Systematic Zoology* **24**, 311-332.

[18] Stephens, J. (1985). Statistical methods of DNA sequence analysis: detection of intragenic recombination or gene conversion. *Mol. Biol. Evol.* **2**, 539-556.

[19] Watterson, G. (1975). On the number of segregating sites in genetical models without recombination. *Theoret. Population Biol.* **7**, 256-276.

[20] Watterson, G. (1989). Allele frequencies in multigene families. II. Coalescent approach. *Theoret. Population Biol.* **35**, 161-179.

[21] Weiss, E., L. Golden, R. Zakut, A. Mellor, K. Fahrner, S. Kvist, and R. Flavell (1983). The DNA sequence of the $H\text{-}2K^b$ gene: evidence for gene conversion as a mechanism for the generation of polymorphism in histocompatibility antigens. *EMBO J.* **2**, 453-462.

[22] Wills, C. (1973). In defense of naive pan-selectionism. *Amer. Naturalist* **107**, 23-34.

ACKNOWLEDGEMENT OF REFEREE SERVICE AND EDITORIAL ASSISTANCE

Andre Adler

Krishna B. Athreya

H. Cohen

Somnath Datta

K. Dietz

Richard T. Durrett

A. Frieze

Robert C. Griffiths

Chris C. Heyde

Ira Longini

William P. McCormick

Peter McCullagh

Michael Phelan

Nancy Reid

Gennady Samorodnisky

D. Sharnbhag

M. Sorensen

J. Michael Steele

G.A. Watterson